Pandemics, Science and Policy

Palgrave Studies in Science, Knowledge and Policy

Series Editors: **Katherine Smith**, University of Edinburgh, UK; **Richard Freeman**, University of Edinburgh, UK

Titles include:

Katherine Smith
BEYOND EVIDENCE-BASED POLICY IN PUBLIC HEALTH
The Interplay of Ideas

Sudeepa Abeysinghe
PANDEMICS, SCIENCE AND POLICY
H1N1 and the World Health Organization

Palgrave Studies in Science, Knowledge and Policy
Series Standing Order ISBN 978–1–137–39461–3 (Hardback)
(*outside North America only*)

You can receive future titles in this series as they are published by placing a standing order. Please contact your bookseller or, in case of difficulty, write to us at the address below with your name and address, the title of the series and the ISBNs quoted above.

Customer Services Department, Macmillan Distribution Ltd, Houndmills, Basingstoke, Hampshire RG21 6XS, England

Pandemics, Science and Policy

H1N1 and the World Health Organization

Sudeepa Abeysinghe
University of Edinburgh, UK

© Sudeepa Abeysinghe 2015

All rights reserved. No reproduction, copy or transmission of this publication may be made without written permission.

No portion of this publication may be reproduced, copied or transmitted save with written permission or in accordance with the provisions of the Copyright, Designs and Patents Act 1988, or under the terms of any licence permitting limited copying issued by the Copyright Licensing Agency, Saffron House, 6–10 Kirby Street, London EC1N 8TS.

Any person who does any unauthorized act in relation to this publication may be liable to criminal prosecution and civil claims for damages.

The author has asserted her right to be identified as the author of this work in accordance with the Copyright, Designs and Patents Act 1988.

First published 2015 by
PALGRAVE MACMILLAN

Palgrave Macmillan in the UK is an imprint of Macmillan Publishers Limited, registered in England, company number 785998, of Houndmills, Basingstoke, Hampshire RG21 6XS.

Palgrave Macmillan in the US is a division of St Martin's Press LLC,
175 Fifth Avenue, New York, NY 10010.

Palgrave Macmillan is the global academic imprint of the above companies and has companies and representatives throughout the world.

Palgrave® and Macmillan® are registered trademarks in the United States, the United Kingdom, Europe and other countries.

ISBN 978–1–137–46719–5

This book is printed on paper suitable for recycling and made from fully managed and sustained forest sources. Logging, pulping and manufacturing processes are expected to conform to the environmental regulations of the country of origin.

A catalogue record for this book is available from the British Library.

A catalog record for this book is available from the Library of Congress.

Contents

1	Introduction	1
2	Narrating the Nature of H1N1	6
3	Risk and Scientific Uncertainty	31
4	Categorizing H1N1 – The Pandemic Alert Phases	64
5	Vaccines, Institutions and Pandemic Management	102
6	Contestation and the Council of Europe	133
7	Globalization and Global Public Health	172
8	Conclusions	204

Appendices	210
Notes	214
Bibliography	215
Index	235

1
Introduction

There has been increasing public health emphasis upon the management of global disease threats. In particular, it has been suggested that, judging from the historical rate of incidence, a severe worldwide influenza pandemic is likely to be imminent (Lazzari & Stohr, 2004; Webby & Webster, 2003; Webster, 1997). Such an event carries the potential to cause widespread social and economic disruption. This risk therefore gives rise to a range of institutional and public expectations and reactions. A climate of heightened vigilance and surveillance, and both pre-emptive and reactionary health measures, result.

Intensified awareness surrounding the pandemic potential of influenza has resulted in a number of global pandemic scares. Prominent examples include SARS (2003) and H5N1 avian influenza (2004–2006). The largest recent global alert surrounded the 2009 A/H1N1 strain of influenza, which is commonly referred to as swine flu. Critically, unlike both SARS and avian influenza, on 11 June 2009, H1N1 was officially declared by the World Health Organization (WHO) to constitute an influenza pandemic – the first pandemic declaration in 40 years (Cohen & Enserink, 2009). From a critical social scientific perspective, this declaration was not merely a result of a set of scientific facts which objectively characterized H1N1 as a 'pandemic'. Rather it was a consequence of socially negotiated definitions of both the H1N1 virus and the term 'pandemic', which was apparent in the discourse and actions of various public health stakeholders. The most notable of these actors was the WHO.

Within the contemporary framework of global public health, the WHO is principally responsible for the monitoring and reporting of infectious disease threats and for organizing and coordinating global reactions. Most importantly, the WHO is also solely responsible for producing authoritative global definitions of the term 'pandemic', and declaring whether any given threat constitutes a pandemic event. In this way, the actions of the WHO, and the conceptions of disease which underlie these actions, are fundamental to the social framing of a disease as 'pandemic', and the global reactions that follow. The WHO's June 2009 declaration of the H1N1 pandemic produced reactions from governments and public health bodies worldwide. It prompted the implementation of national pandemic preparedness plans and global reactions, such as the production and distribution of vaccines and a heightened interest in border control.

The H1N1 virus spread globally and at a rapid rate following its initial detection (refer to Appendix 1 for a timeline of events). However, as the situation developed, it became increasingly clear that the 2009 strain would not result in high morbidity and mortality. By the WHO's official declaration of the end of the pandemic on 10 August 2010, only approximately 18,500 laboratory-confirmed deaths had resulted from H1N1 globally (WHO Situation Update, 11 August 2010). Though the measurement of mortality in the case of pandemics is difficult to quantify (Monto, 1987), it is clear that in relation to previous influenza pandemics, which produced death rates from approximately 33,800 in the USA and 30,000 in England and Wales for the least severe (Hong Kong Influenza, H3N2, 1968/1969) through to 50 million globally for the most severe (Spanish Influenza, 1918/1919), the H1N1 pandemic was comparatively mild (Cox & Subbarao, 2000; Nguyen-Van-Tam & Hampson, 2003; Taubenberger & Morens, 2006).

As a reaction to a perceived lack of impact, the pandemic declaration by the WHO, and the actions which followed it, were called into question by numerous state and public bodies. These actors questioned fundamental facets of the WHO's construction, including the organization's characterization of H1N1, its definition of the concept of 'pandemic' and its depiction of risk. First and foremost among the institutional critics was the Council of Europe, which projected the concerns of European member states with regard to the WHO's management of H1N1. The ensuing controversy highlighted the centrality of the WHO's construction of the threat in framing reactions, and the fragile nature of those constructions.

The evidence that the WHO's perspective was susceptible to criticism shows that the its construction of H1N1 had not obtained scientific closure. In fact, the WHO's depiction of the H1N1 pandemic was fundamentally unstable, rendering the critique of its response possible. The case study of the H1N1 pandemic therefore demonstrates the centrality of the social construction of scientific fact in framing the perception and management of infectious disease threats. It also demonstrates how the accounts of the actor responsible for defining the 'fact' of infectious disease (here the WHO) can become contested. This contestation was a consequence of the lack of closure and inherent ambiguity in the underlying construction of the phenomenon.

In attempting to understand the contestation of the WHO's management of H1N1, two important questions emerge: How was the disease constructed by the WHO in such a way as to precipitate global action, and how was this construction rendered liable to fundamental critique? This book seeks to understand how the H1N1 pandemic was constructed and managed by the key defining organization of the WHO. It furthermore aims to explain the mechanisms which rendered those constructions and management strategies vulnerable to critique by outside actors. In doing this, it investigates the way in which the WHO represented the H1N1 pandemic, including the organization's risk narrative surrounding the event. Second, the book explores the wider social and institutional structures which formed the WHO's account and subsequent management of the disease. Third, given that the WHO's perspective became widely contested, the book seeks to understand the lack of scientific closure surrounding the concept of the H1N1 pandemic. It investigates why the WHO's construction was fragile, and demonstrates how this led to the contestation of the WHO's account by the prominent critic of the Council of Europe.

Through an analysis of statements and documents from the time of the pandemic, the book investigates the way in which the WHO conceptualized and constructed both the specific infectious agent, influenza A/H1N1, and the notion of 'pandemic'. It focuses upon the following key questions:

- How did the WHO represent the nature of H1N1?
- How did the WHO characterize H1N1 as a 'pandemic'?
- How did the WHO represent the risk surrounding H1N1?
- How did the WHO characterize its reactions to H1N1?
- What institutional structures underpinned the WHO's representation and management of H1N1?
- What other social factors played a part in producing the WHO's representation and management of H1N1?
- Given the contestation of the WHO's account, in what ways had it been rendered susceptible to contestation?

- What was the basis of the Council of Europe's contestation of H1N1, and in what ways did the WHO's representation determine the substance and form of this critique?
- Thus, as this work progresses I will explore the characteristics of the WHO's representation of the H1N1 pandemic, and the way in which this representation became open to contestation.

Embedded within a context of scientific uncertainty, and following an institutionalized reaction to infectious disease and a reframing of roles within global public health, the WHO's construction of the H1N1 pandemic was rendered liable to significant external critique. As I examine each aspect of the WHO's management of H1N1 in turn, I show that that the WHO's framing of H1N1 as a pandemic threat was fragile and unstable as a result of the context of scientific uncertainty, institutional path dependence and shifting institutional roles within global health. Combined with the perceived mildness of the disease as events unfolded, and the democratized nature of scientific research, the WHO's account became susceptible to contestation by outside actors.

I take a look at the problem of H1N1 in a holistic manner, starting from the small-scale characterization of the problem (the definition of the virus itself) and enlarging my focus to look at the problem of H1N1 in relation to the structures of global public health. I first explore the WHO's construction of the H1N1 virus (Chapter 2), which is key to framing actors' reactions to the pandemic. I investigate the inherent fragility of that construction, and examine the ways in which this uncertainty underpinned subsequent events. I move on to studying the way in which the WHO framed the problem of pandemic risk, and show how the organization attempted to maintain this characterization of risk despite the evident mildness of the disease over time (Chapter 3). Next (Chapter 4) I explore the fact that this risk construction was only possible through the institutional definition and classification of pandemic threats, made through the WHO's Pandemic Alert Phases. As such, I investigate these phases, looking at their definitional ambiguity, and demonstrating that the WHO's classification of 'pandemic' was ill-constructed and, combined with the lack of disease severity (and with the fragility of the initial construction of H1N1), was liable to outside critique.

Moving away from the act of defining the pandemic, I broaden my gaze to examine the institutional processes and politics that underpinned the response to H1N1 (Chapter 5). I start by examining the WHO's reaction. Since the WHO had depicted H1N1 as a high risk, it needed to take some action in its management. I demonstrate how this action was framed through path-dependent institutional processes, which led to a significant emphasis upon mass vaccination. The WHO's characterization and actions surrounding H1N1 were formed within conditions of scientific uncertainty and (path-dependent) entrenched institutional process. This resulted in the contestation of the organization's decision-making by many outside actors.

Looking at these criticisms sheds further light on the WHO's construction and management of H1N1. An important voice among these was that of the Council of Europe, whose critique of the WHO I explore next (in Chapter 6). Here, the fragility of the WHO's construction of H1N1 and 'pandemic' come to the forefront, and this is fundamental to the Council of Europe's critique of the institution. To end, (Chapter 7) I broaden my focus even further – to the effects and impacts of this event on the structures and institutions of global public health. I demonstrate that the instability of the WHO's constructions, and the ability of the Council of Europe to question them, were framed within the wider structure of global public health. Here I show that the changing nature of public health undermined the authority of the WHO and in part helped to produce the lack of clarity and closure in the its construction of the H1N1 pandemic.

It is clear that the instability of the WHO's construction of H1N1 was a function of the social context within which the organization was acting. This book demonstrates the manner in which the definition of a pandemic can become fundamentally open to contestation. It illustrates the impact of scientific uncertainty on the management of contemporary global risks, contributing to the understanding of scientific knowledge production under conditions of uncertainty. The case study of the WHO's management of H1N1 therefore helps to illuminate both the contemporary reaction to pandemics and the problems of risk-managing institutions in dealing with fundamentally uncertain and novel events.

2
Narrating the Nature of H1N1

For a scientific object or idea to be accepted by all actors who engage with it, it first needs to reach stability as an incontestable 'fact'. In the case of H1N1, the institution responsible for this fact-making was the WHO, since it is accountable for defining and managing global disease threats. Throughout this book I will examine how the WHO failed to effectively mobilize a stable construction of the 'H1N1 pandemic', ultimately resulting in the contestation of the concept by prominent global health actors. However, in order to explore the reasons why the H1N1 pandemic proved to be a fragile concept, it is first necessary to illustrate the elements of the WHO's initial construction. This chapter examines the WHO's attempt to define the phenomenon of H1N1. It argues that there were several factors of the construction that lent to its fragility as a scientific fact. These include a lack of early consensus on the name, a failure to articulate a robust and coherent origin narrative, and ineffectual comparisons with seasonal influenza and historical pandemics. These inadequacies meant that the concept of H1N1 did not reach definitional 'closure' (Callon, 1986; Pinch and Bijker, 1984), rendering it open to contestation.

In explaining the importance of constructing a stable notion of the H1N1 pandemic, the concept of 'translation' from actor-network theory (ANT) is valuable. ANT understands social life as consisting of associations of actor networks, such that each social actor is in fact underpinned by a complex network of other actors, as well as relationships between actors. What we understand to be the actor is rather a 'punctualization' – it is the condensation of an actor network onto one point (Latour, 1996; Law, 1992; Prout, 1996). The process through which these actor (network)s are formed, and punctualization is stabilized, is known as 'translation'.

As first developed by Callon (1986), the term 'translation' refers to the process by which actors (including concepts/'things', such as a H1N1 pandemic in this case) come into existence. Translation occurs in four stages. The first is problematization, where the actor network (i.e. the 'thing' in question) is initially built. This stage is the subject of Chapter 3, which will explore the way in which the network-building agent (i.e. the WHO) attempted to build a stable actor network for the concept 'H1N1 pandemic'. The second and third stages are 'interessement' and 'enrolment', which refer, respectively, to the representation of the new 'thing' to outside actors and the enrolment of other actors into an association with the thing. Although the successful enrolment of actors such as national governments was necessary to the translation of the pandemic (see Chapter 6 for a discussion of the WHO's failure in enrolling outside actors in to the H1N1 actor network), this did not occur. In the case of H1N1, these processes were not achieved effectively because of the fragility of the initial problematization. The fourth and final stage – 'mobilization' – should occur when the developed actor network (H1N1 pandemic) is stabilized and can act in an uncontested manner – by now, effective punctualization has been performed (Callon, 1986; Latour, 2005; Law, 1992). In the case of HN1, as will be furthered argued in subsequent chapters of this book, mobilization was ineffective. As a result, the H1N1 pandemic became a questioned and contested concept.

The WHO narrative failed to demonstrate that H1N1 fulfilled the characteristics of a 'pandemic'. Furthermore, it failed to effectively distinguish H1N1 from 'non-pandemic' disease. Thus the initial problematization of H1N1 was not conducted successfully. This meant that the WHO did not mobilize an effective translation of the 'H1N1 pandemic', a fact which eventually led to the breakdown and contestation of the concept as a whole.

What/when is a pandemic?

To appreciate the WHO's attempts to translate H1N1 as a pandemic threat, it is necessary to understand the way in which the organization depicted the general category of 'pandemic'. Prior to the controversy and criticism surrounding H1N1, the concept was treated by the WHO (and other health authorities) as unproblematic. It was taken for granted that a true pandemic could be distinguished as such; if a pandemic event occurred, it would be easily and clearly discerned. In short, using the terminology of ANT, the concept of 'pandemic' was well-punctualized and 'black-boxed'; the network behind the actor 'pandemic' had been rendered invisible, and was indisposed to investigation or 'opening' by other actors which interacted with it (Latour, 1987). Prior to H1N1, the term 'pandemic' was utilized unproblematically, underpinned by the implicit assumption that all relevant actors understood what constitutes a pandemic. Since the concept was taken to be unproblematic, this black-boxed conceptualization was the frame through which early reports of H1N1 were viewed by the WHO.

The concept of 'pandemic' became increasingly problematic as the case of H1N1 developed. Through the initial problematization of the 'H1N1 pandemic' actor network, the concept was necessarily associated (i.e. was networked) with the broader idea of 'pandemic'. When the actor network of 'H1N1 pandemic' failed to be effectively mobilized, the formerly unproblematic concept was also rendered liable to critique through association (see Chapter 4). However, during the initial construction of H1N1, a black-boxed understanding of 'pandemic' was used in referring to H1N1, in the attempts to problematize it as a pandemic. To appreciate the subsequent contestation of both concepts (H1N1 and pandemic), I will investigate these early uses.

In the early WHO documentation, a pandemic was depicted as characterized by a number of distinct features. These were the novelty of the pandemic agent; the unpredictability of the virus; the ability of the virus to spread quickly over a large geographical region; the ability of the virus to swiftly mutate into different forms; the mass susceptibility of global populations to the virus; and a differentiation from seasonal influenza. In part, these assumptions were ineffectively articulated by the WHO's representatives. This is because the black-boxed nature of the concept 'pandemic' rendered it difficult for actors to propound a definition, since meaning was assumed. However, given the problematic nature of H1N1, the WHO's representatives were forced to verbalize their understanding of the concept 'pandemic'. The way in which the representatives described these aspects of 'pandemic' will be discussed briefly in turn to show how these features were mobilized. Many of these themes will be discussed in further detail throughout the book. At this stage, though, it is important to illustrate the WHO's depiction of the nature of pandemics. Ultimately, this demonstrates the inadequacy of the WHO's assemblage of the term 'pandemic' in the translation of H1N1.

When pressed to explain the nature of pandemics, the initial black-boxed understanding of 'pandemic' was depicted by the WHO as comprising a few interacting variables. First, the WHO heavily emphasized the novelty of a viral strain in characterizing a pandemic agent. Viral novelty links with immunological susceptibility, since (along with other factors) a virus with unique antigenic properties suggests the capacity to uniquely challenge pre-existing immunities to influenza (Cannell et al., 2008; Mathews et al., 2009). In this way, one of the proposed characterizing features of a potential pandemic strain was novelty. Thus

> Influenza pandemics are caused by a virus that is either entirely new or not known to have circulated among humans in recent decades. This means, in effect, that nearly everyone in the world is susceptible to infection. It is this almost universal vulnerability to infection that makes influenza pandemics so disruptive.
>
> (Chan, 04/05/09 [Please refer to appendix 2 for role of key actors])

In reference to H1N1, it was suggested by the WHO that 'This particular H1N1 strain has not previously circulated in humans. The virus is entirely new' (Chan, 17/06/09). For the WHO, the disruption of a pandemic was a function of the novelty of the infectious agent. In this way, it was made clear by the organization that novel viral agents must be monitored closely due to their capacity to spread globally, and that this potential characterized the threat that is posed by a pandemic.

The notion of spread was therefore also fundamental. It was asserted by the WHO that a large (global) geographical spread is another characterizing feature of a 'pandemic'; in many contexts, it was suggested to be the defining feature. Thus, for example, Fukuda suggested that

> An easy way to think about pandemic – and actually a way I have sometimes described in the past – is to say: a pandemic is a global outbreak. Then you might ask yourself 'What is a global outbreak?' Global outbreak means that we see both spread of the agent – and in this case we see this new A(H1N1) virus to most parts of the world – and then we see disease activities in addition to the spread of the virus.
>
> (Fukuda, 26/05/09)

The ability to spread quickly was therefore assumed to be a notable characteristic of pandemic influenza, and disease activities are presumed to be a function of spread. Thus it was suggested that 'Influenza pandemics must be taken seriously precisely because of their capacity to spread rapidly to every country in the world' (Chan, 29/04/09), and (following the declaration of H1N1 as a pandemic, and the wide spread of the virus) that 'As we see today with well over 100 countries reporting cases, once a fully fit pandemic virus emerges, its further international spread is unstoppable' (Chan, 17/08/09). The global and 'unstoppable' spread of a pandemic strain again served as evidence that warranted concern regarding the threat, and defined H1N1 as a 'pandemic' in the WHO's account. The supposedly defining feature of 'spread' was used to translate 'pandemic' into the concept of 'H1N1 pandemic'.

Also linked to the concept of novelty and spread was the unpredictability of a potential pandemic agent. Thus it was suggested that 'New diseases are, by definition, poorly understood. Influenza viruses are notorious for their rapid and unpredictable behaviour' (Chan, 29/04/09). This defining quality of unpredictability was emphasized and reiterated on numerous occasions. In this way it was asserted that

> Influenza viruses are the ultimate moving target. Their behaviour is notoriously unpredictable. The behaviour of pandemics is as unpredictable as the viruses that cause them. No one can say how the present situation will evolve.
>
> (Chan, 11/06/09)

And again,

> this early, patchy picture can change very quickly. The virus writes the rules and this one, like all influenza viruses, can change the rules, without rhyme or reason, at any time.
>
> (Chan, 17/06/09)

The unpredictable genetic mutability of the influenza virus was noted as a point of concern, potentially producing uncertain health outcomes. However, the unpredictable contribution of other external variables also served to necessitate concern about pandemic disease. Thus, 'Apart from the intrinsic mutability of influenza viruses, other factors could alter the severity of current disease patterns, though in completely unknowable ways' (Chan, 15/05/09). In this way the concept of unpredictability in and of itself was fundamental to the definition of 'pandemic'.

Simultaneously, through the WHO narrative, human agency was portrayed as the ultimate means by which to combat infectious disease. Thus a recurring theme throughout the director-general's speeches reflected that although 'influenza viruses have the great advantage of surprise on their side', they are combatable because 'viruses are not smart. We are' (Chan, 11/06/09). To some extent, then, a pandemic state was defined by the unpredictability of human reactions as well as the unpredictability of viral agents, and although the virus cannot be easily managed, the global response can. More practically, statements such as the above also serve to reinforce the necessity of the WHO as an institution to manage global pandemics, as I will explore in Chapter 7.

In addition to uncertainty, the susceptibility of global populations in another important feature is the black-boxed definitions of 'pandemic'. It was suggested, for example, that

> Influenza pandemics are remarkable events because they spread throughout a world population that is either largely or entirely susceptible to infection. They tend to hit a given area in the epidemiological equivalent of a tidal wave.
>
> (Chan, 17/08/09)

This capacity for spread through a global population was, for the WHO, at the heart of the effects of a pandemic:

> if we do move into a pandemic, then our expectation is that we will see a large number of people infected worldwide. This is typically what happens in pandemic situations. If you look at past pandemics, it would be a reasonable estimate to say that perhaps a third of the world's population would get infected with this virus. You never know beforehand, but this would be a reasonable kind of estimate. When you look at a third of the world's population – in recognizing that we are a globe of a little over six billion people – that is a lot of people to get infected.
>
> (Fukuda, 07/05/09)

The idea of a third of global infection was retracted almost immediately after it was made, but this emphasis on geographical spread was fundamental to the WHO's account. As seen in the quote above, the organization also implicitly presumed a high correlation between geographical spread and morbidity rates. This assumption of severity was clearly black-boxed and, early in the WHO's account, so implicit as to not even have been narrated. As I will show in chapters 3 and 4, one of the factors which served to disrupt the punctualization of 'pandemic' was that, in the case of H1N1, global geographical spread occurred but morbidity rates did not mirror this. However, initially, susceptibility and therefore presumed morbidity underpinned the black-boxed depiction of a pandemic state, which H1N1 was translated into.

The spread of pandemic influenza and the susceptibility of the global population was also implicated in another characterizing feature of a pandemic as described by the WHO. This was, simply put, that pandemic influenza is fundamentally different from seasonal influenza. Wide geographic spread and susceptibility were narrated as atypical of non-pandemic (seasonal) influenza viruses. This helped to justify the WHO's pandemic declaration. The WHO suggested that pandemic epidemiological patterns are

> not what we see with normal seasonal influenza. When you put all of these things together, what it really suggests is that we are in a situation which is really moving towards more or less a pandemic type of situation. The pandemic really refers to the fact that we are seeing the geographical spread of a virus that is causing this disease.
> (Fukuda, 09/06/09)

The WHO's effort to distinguish pandemic from seasonal strains was an important theme as the definition of H1N1 as pandemic became increasingly fragile through the course of events.

Finally, in depicting a pandemic, it was also suggested by the WHO that it is a long-term rather than an acute event. Here it was asserted that previous pandemics were not immediately apparent in their beginnings. In this regard,

> Pandemics do not occur in a couple of days. When we go back and we look at history – we are often looking at a one-year period – but really if you look over a two-year period that is really the period in which we see an increase in the number of illnesses and deaths during pandemic influenza.
> (Fukuda, 07/05/09)

As with the assertions of risk and 'evolution' illustrated in Chapter 3, this suggestion reinforced the argument that long-term monitoring and action are necessary against a pandemic event.

Thus, in the initial definition of 'pandemic', the designation of a pandemic threat relied on a number of seemingly well-defined and unproblematic characteristics. These included novelty, spread, susceptibility and unpredictability. Prior to the H1N1 case, these aspects were taken by the WHO to be objectively and unmistakably observable and/or quantifiable. However, as I will show, these components of definition can all to some extent be rendered fragile and tenuous. As the H1N1 pandemic (and the controversy surrounding it) developed, the concept of pandemic itself was reproblematized and the terms of definition became destabilized. (The resulting re-evaluation of these terms will be illustrated as the book progresses.) For a stable problematization of H1N1 to have occurred, the translation of the event as a pandemic at its core relied upon a shared understanding of the characteristics of this category of disease. In principle, the declaration of a pandemic by the WHO occurred at the point when the threat conformed to the defining aspects above, as set out through the Pandemic Alert Phases, and the WHO's definitional document. However, as I will show through this analysis of the H1N1 case, assumed definitional underpinnings can become subject to social contestation and renegotiation, and have done so in relation to the H1N1 'pandemic'.

What is the H1N1 pandemic?

To create the object 'H1N1 pandemic', the WHO needed to stably subsume the broader concept of 'pandemic' into the emerging actor network. To do this it propounded narratives corresponding to the nature of the H1N1 virus with the assumed definitional features of a 'pandemic'. Also important was the way in which the WHO attempted to translate the concept of a pandemic into the emerging phenomenon of H1N1. This also illustrates the inclusion of other concepts which served as attempts to distinguish H1N1 as a novel and distinct infectious agent. Here I demonstrate that attempts at stably translating an H1N1 pandemic were performed through the inclusion of various concepts into the emerging actor network, including the construction of a distinction between H1N1 and seasonal influenza, and the construction of an analogy between H1N1 and historical examples of influenza pandemics. I will argue, however, that this effort at translation was ineffectual. The concepts were not neatly adopted, and this ultimately resulted in the challenging of the construction 'H1N1 pandemic'.

Comparison with seasonal influenza

Necessarily, in order to be defined as a pandemic, an influenza virus strain must be distinguished from seasonal influenza. This distinction was vital for the WHO to mobilize H1N1 as a pandemic strain. This act of distancing is particularly interesting in the case of H1N1 given, as I will examine further, the relative mildness of the disease.

From the beginning, the WHO's attempts to distinguish H1N1 from seasonal influenza were questioned. This is part of the explanation as to why H1N1 failed to be effectively mobilized and recognized as a pandemic. As early as 5 May 2009, press questions focused upon the lack of distinction between H1N1 and seasonal influenza. The WHO's responses to these questions were often ambiguous and failed to provide a clear distinction. For example, in reply to one question which suggested that seasonal flu deaths are in fact potentially large (minimizing the distinction with H1N1), it was stated that

> In fact the numbers we have for seasonal flu vary depending on the years. Some years we have a very mild seasonal flu, and other years we have a more severe seasonal flu. Global figures are really difficult to get because each country is monitoring the seasonal flu, and they provide their figures, but not necessarily on a regular basis. But to give you a kind of frame, in France for example, the number of deaths during seasonal flu varies from 5000–15,000 deaths, in the United States you can have 40,000 deaths depending on the years, so these are numbers, but highly variable.
>
> (Briand, 08/05/09)

The representatives potentially needed to maintain the perception of the impact of seasonal influenza (even within discussions of pandemic strains) given that, on the global scale, seasonal strains do in fact represent a significant health burden and remain an important aspect of the WHO's non-crisis health governance (WHO, 2011a; WHO, 2011b). However, in the context of the discussion of H1N1, the failure to downplay seasonal influenza, or at least to have depicted a strong distinction between pandemic and seasonal strains, constituted a point at which the WHO's problematization of H1N1 became vulnerable. As the quote above shows, in order to make distinctions, the representative appears to have implied that the mortality from seasonal flu can vary, whereas the mortality from pandemic influenza is always high. However, this does not represent a clear-cut marker of difference and fails to construct H1N1 as a distinct event. This was particularly apparent when the events of H1N1 eventuated as mild (as we will see in chapters 3 and 4).

The inadequacy of the attempt at differentiation can be illustrated further by the fact that the H1N1 pandemic did not develop into a high-morbidity threat. However, at the time immediately following the appearance of H1N1, it was assumed by the WHO that the virus would follow the epidemiology of prior pandemic cases. Using the statistics from Briand's quote above, it could indeed be suggested that seasonal strains are of great threat. However, it was also asserted that

> Yes, it is true that seasonal influenza viruses kill people every year. Although there are not very precise estimates for the world, it has been estimated that up to about half a million people per year can die from seasonal influenza infections. Now the reason we are paying so much attention to this virus though, is that seasonal influenza viruses have been around the world and have been circulating for many years. And so we understand their behaviour and we know that most people... have some immunity to them; that is what makes them seasonal influenza viruses. But we also know that when a new influenza virus enters the human population, and people do not have immunity to this virus, then the levels of serious illness and the levels of deaths can be higher than we see with regular seasonal influenza.
>
> (Fukuda, 05/05/09)

The second part of this quote again demonstrates the WHO's argument that the novelty and unpredictability of the H1N1 virus served to distinguish it from seasonal influenza, asserting that the novelty presented a primary source of the threat of a pandemic.

However, the WHO's explanation for the difference between H1N1 and seasonal influenza was unconvincing – there was not a significant distinction made between the two. The inability to clearly distinguish H1N1 from seasonal influenza was evident in early press conferences. Here, direct comparisons between the two states of influenza (pandemic and seasonal) were made in a way which failed to establish a distinction. For example, in one conference it was suggested that with respect to H1N1,

> The illness that we are seeing is generally consistent with seasonal influenza infection. That is the kind of symptoms that the milder cases are experiencing and generally what are seen with other influenza viruses, although there is some suggestion that perhaps cases are developing diarrhoea more often than is normal...
>
> (Fukuda, 29/04/09)

This did not provide a convincing argument that H1N1 is indeed distinct. This lack of distinction was apparent throughout the narrative of H1N1. For example, in another circumstance it was stated that

In terms of the illness itself, in the people who are developing generally milder illness, this is similar to the kinds of influenza-like illnesses that we see... and this is generally in keeping with what the milder spectrum of illness is.

(Fukuda, 05/05/09)

Thus, early characterizations of H1N1 failed to establish a clear difference between the purported pandemic and seasonal strains. The impact of H1N1 was not clearly distinguished from seasonal influenza, thereby contradicting its attempted problematization as a unique and separate entity. This lack of distinction was therefore detrimental to the effective translation of the concept of the H1N1 pandemic.

Of perhaps even greater importance is the fact that this early failure to create distinctions was not remedied as the pandemic progressed. In fact, while comparisons with seasonal influenza were made at the early stages of the threat, later on, direct comparisons with seasonal influenza were not made by the WHO's representatives, and questions of that nature were deflected onto an emphasis upon geographical spread.

Historical analogy

As demonstrated above, one of the ways in which the WHO attempted to translate the novel phenomenon of a H1N1 pandemic was to distinguish it from seasonal influenza. A second important part of the translation involved comparison with past pandemic threats, emphasizing the similarity to past experience. In general, drawing upon the collective memory of past contagion is fundamental to the construction of a new disease. This is because the characterization of any new threat necessarily reflects pre-existing conceptualizations of infectious disease (Fleck, 1979), and one of the ways in which a novel phenomenon is understood is through reference to an existing comparative framework (Marková & Farr, 1995). Since such analogy construction constitutes an important device through which thoughts and ideas are represented and analysed (Arber, 1954; Sontag, 1978), the translation of the H1N1 pandemic actor network necessarily involved the assembly of these links. The adoption and acceptance of such an analogy was crucial to the problematization of the H1N1 concept as a whole. The WHO was more successful in the use of historical analogy than in its attempts to distinguish H1N1 from seasonal strains. However, the acceptance of these links by other actors was still not unquestioned, and conceptual links were weak.

From the very first press conferences on the event, the H1N1 virus was introduced by the WHO's representatives in relation to previous infectious diseases. In one of the opening statements of the first conference, attempts at analogy with previous pandemics was evident:

> Many of you know that the world has been talking about and preparing for pandemic influenza for at least the past five years and there are a number of reasons for this. We know that influenza pandemics have occurred at least a couple of times each century and in the last five years we have been working very hard...because of a specific pandemic threat known as avian influenza or H5N1 and because of that many countries have been very focused on strengthening their defences for such a situation.
>
> (Fukuda, 26/04/09)

In this statement it was asserted that influenza pandemics are an ever-present risk that should be (and have been) prepared for. The initial problematization of the H1N1 threat was therefore situated in the context of the wider historical narrative of the frequency and effect of pandemic disease.

Such reminders of the possibility of a pandemic recurred throughout early WHO statements. It was suggested that 'we have seen such occurrences a couple of times each century and so the question right now is whether we are in such a situation...' (Fukuda, 26/04/09). In a specific analogy between the present threat of disease and past experiences, it was stated that

> the world continues to be threaten[ed] by these new emerging infections. This is not something of the past, this is an ever present reality for the world. Ever since from SARS, the introduction of AIDS in that time period up till now, there will be any number of a new important emerging infectious diseases of which SARS and avian influenza have been some of the most important...
>
> (Fukuda, 28/04/09)

Thus, one of the ways in which analogy to past pandemics was mobilized was through suggestions of the ubiquity of pandemic threats more generally, including analogy even to threats which were biologically quite different, such as the AIDS pandemic. However, analogous respiratory diseases (i.e. avian influenza and SARS) were depicted as of primary concern.

Reference to past pandemics also served to reinforce the unpredictability of pandemic influenza, suggesting that the world needed to be vigilant about the volatility of H1N1's pandemic potential. Thus 'experience of past pandemics warns us that the initial situation can change in many ways, with many, many surprises' (Chan, 04/05/09). These types of analogy reinforced and justified concern about H1N1 despite what was at the time the mild epidemiological presentation of illness (when it was as yet uncertain how the pandemic would unfold). Here, past pandemics were specifically invoked to reinforce the potential impact of the virus, as illustrated in the following example.

> the 1918 pandemic, the most deadly of them all, began in a mild wave and then returned in a far more deadly one. In fact, the first wave was so mild that its significance as a warning signal was missed.
>
> ... the pandemic of 1957 began with a mild phase followed, in several countries, by a second wave of greater fatality. The pandemic of 1968 remained, in most countries, comparatively mild in both its first and second waves.
>
> At this point, we have no indication that we are facing a situation similar to that seen in 1918. As I must stress repeatedly, this situation can change, not because we are overestimating or underestimating the situation, but simply because influenza viruses are constantly changing in unpredictable ways.
>
> (Chan, 04/05/09)

The above quote demonstrates the WHO's attempt to reinforce the concept that there was a lack of initial severity in the three major past pandemics, suggesting that H1N1 could mimic these events and eventually manifest as a severe disease. Furthermore, such assertions served to construct the characteristic of unpredictability: H1N1 was represented as threatening due to its unknown impact.

The quote above also makes reference to Spanish influenza as the prototypical example of severe pandemic. Spanish influenza was an important framing device in translating the threat of H1N1. For example, it was again suggested that

> the worst pandemic at the last century started out mild in the springtime, it was fairly quiet during the summer, and then in the autumn when it really exploded, this is in 1918 and it was a much more severe form.
>
> (Fukuda, 29/04/09)

Such assertions recount the need for constant vigilance and emphasize the potential threat of the current situation. The inclusion of the concept of the Spanish influenza pandemic was an important device in producing a stable translation of H1N1 due to the prominence of the example of Spanish influenza within the collective understanding of pandemics. The Spanish influenza pandemic of 1918–1919 is upheld as a prototypical example of an influenza pandemic despite its epidemiological uniqueness (Taubenberger & Morens, 2006; Tognotti, 2003), and it distinguishes itself as a fearful event in public memory (Barry, 2004; Crosby, 1976). Spanish influenza is therefore an important device in invoking public memory of pandemic disease (Abeysinghe & White, 2010). To conduct the problematization of H1N1 as a legitimate pandemic threat, the WHO therefore utilized the device of analogy to portray H1N1's pandemic nature, constructing and reinforcing links between Spanish influenza and H1N1.

However, it is equally notable that, while the WHO can be seen to have attempted to mobilize an analogy with Spanish influenza, at points it was suggested by other actors that the 1918 disease conditions differed significantly with those of potential contemporary pandemics. For example, one reporter questioned whether 'given the improvement in today's medicine is there really a chance of a repeat if 1918 or is that really not a possibility?' (Fukuda, 05/05/09). In response it was suggested that

> our medical technology is just better than it was back in 1918, we have better antibiotics, we have better medical care than we had, can we really expect to see at some point a repeat of that phenomenon? Hopefully not. But I think that one of the realities of these kinds of global health events, is that the current systems that we have can also be overwhelmed. When we went through the SARS epidemic a few years ago, in the places that were heavily affected we saw that the medical capacities could be overwhelmed easily.
>
> (Fukuda, 05/05/09)

It was thus argued by the WHO that, while the world was more prepared, catastrophe was still possible given the potential impact of infectious disease threats. In this way, this exchange again served to reinforce the unpredictability and potentially severe impact of pandemics even in the contemporary climate.

In another example, the representative reacted to studies which suggest that the 2009 H1N1 strain was not as virulent as that which caused Spanish influenza:

> There are in fact some characteristics of the virus, I mean they have identified some genes that are more virulent and can give more

virulence to the virus itself. Especially people are often comparing this virus to the 1918 pandemic virus as a kind of standard for comparison. *It seems in fact that this virus does not have this kind of characteristic.* However, as I said before, this is not enough to say that it will be mild because first of all, apparently, it is quite a new virus so most of the population is completely naïve to this virus.

(Fukuda, 13/05/09; emphasis added)

Here the WHO was responding to a study which suggested that the biological characteristics of H1N1 indicate a milder strain. However, in the face of this evidence, it was again asserted that the overall impact of the spread of the virus might be devastating. To maintain this argument, the speaker made links to the novelty of the virus and the susceptibility of global populations. However, suggestions such as the study which this quote alludes to also served to undermine the problematization of the new H1N1 actor network, as they highlighted some apparent contradictions in its construction.

Nevertheless, allusion to past pandemics was an important network-building strategy. References to other past events were made in order to attempt to problematize that threat and to counteract the suggestions of mildness which could undermine the problematization. As will be demonstrated in Chapter 3, the notion of severity became a highly contested concept in the case of H1N1. In terms of the use of historical analogy, throughout the early conferences it was suggested that it was difficult to gauge the severity of pandemics, and that historical examples illustrate this point. Thus,

We are working with disease modellers, we are working with epidemiologists to get a better handle on that but I think if you cast your mind back to SARS, if you cast your mind back to other epidemics, at this stage in an epidemic it is sometimes very difficult to make an accurate estimate of severity.

(Ryan, 02/05/09)

Moreover, references to history were also used to assert that pandemics can range in severity. This technique was increasingly prominent in later stages of the pandemic development, when it became clear that H1N1 was unlikely to ever produce severe disease. Severity was linked to the more clearly defined characteristics of disease, such as geographical spread and novelty. Thus it was argued by the WHO that

> In the past when pandemics have occurred...you have a new influenza virus arrive, [and] begin to spread around the world because it is being transmissible among people. We have seen pandemics cause relatively fewer deaths, and fewer serious illness, this was true in 1968. And we have also seen pandemics cause huge numbers of deaths. In 1918 the most conservative estimates of death, in that one year period, ranged from about 20 to 40 million people dying in one year from that infection. And we also know in that pandemic, it started out mild in the spring time and then over the course of several months became a severe illness.
>
> (Fukuda, 05/05/09)

This quote illustrates attempts to problematize H1N1 in two ways. First, it suggested that pandemics may be mild (such as 1968) while still maintaining the ontological status of the 'pandemic'. Second, it asserted that events which are initially experienced as mild might rapidly become severe (such as 1918). In this way, reference to past pandemics served to construct H1N1 as a pandemic and to reinforce the potential impact of its spread.

The WHO also attempted to define H1N1 as a pandemic in other ways. In addition to historical analogy, references to concurrent disease events were also made. Linking H1N1 to past threats served to place it into the context of a continuing history of disease. In part, analogy to past pandemics was also used to assert that this experience 'helps us to understand the situation, right now' (Chan, 04/05/09); it was suggested that H1N1 can be recognized/understood through this accumulated knowledge. Furthermore, as a result of past pandemics, preparation for pandemic events has occurred so that 'the world today is much more alert to such warning signals [of the appearance of new strains] and much better prepared to respond' (Chan, 04/05/09).

However, to effectively problematize H1N1 as a current threat, the WHO needed to situate the emerging actor network within the contemporary infectious disease climate. References to more contemporary threats served a different purpose from historical analogy – such allusions represent attempts to translate H1N1 as a significant actor network within contemporary global disease.

The H5N1 (avian) virus was related to H1N1 throughout. For example, the impact of an H1N1 pandemic upon avian influenza was portrayed as a concern since 'No-one can predict how the H5N1 virus will behave under the pressure of a pandemic' (Chan, 17/05/09). Furthermore, in the

WHO's account, the rapid transmission of H1N1 was used to represent the 2009 pandemic as a more pressing concern in comparison with avian influenza. It was stated that,

> Unlike the avian virus, the new H1N1 virus spreads very easily from person to person, spreads rapidly within a country once it establishes itself, and is spreading rapidly to new countries.
>
> (Chan, 11/06/09)

Again, this reinforced the suggestion that transmissibility and spread superseded any emphasis upon the concept of severity in the WHO's account (see Chapter 4). Overall, allusions to avian influenza helped to problematize H1N1 as a threat not only in itself but also in its impact upon existing patterns of disease. This was an important part of the problematization because it marked the embedding of H1N1 into the existing disease environment. This served to reinforce the construction of H1N1 as a significant contemporary global health actor.

Thus an important component of the WHO's translation of the concept of an H1N1 pandemic was its use of analogy to historical and contemporary disease. Such comparisons served to highlight the threat of H1N1, construct notions of mildness and severity, and provide indicators regarding justified preparatory measures.

Origins and zoonosis

To successfully translate a new disease, in addition to relating it to previous incidences of illness and the contemporary health environment, it is also necessary to produce a coherent origin narrative. Disease narratives are important in the translation of a disease because they help to socially locate the threat and give meaning to disorder (Douglas, 1969; Nelkin & Gilman, 1991). Sociologically speaking, explanations of origin are an important component of infectious disease narratives because understanding origins helps to make sense of the experience of disease (Herzlich & Pierret, 1987). From an actor-network perspective, origin stories are useful to the successful translation of a new disease in allowing for a successful initial problematization by accounting for the appearance of the new 'thing' on the social landscape. Given this importance of explaining the initial source of disease, it is notable that the WHO did not present a coherent origin narrative in the initial period of translation. The lack of an origin narrative was a deficiency in the WHO's translation of the concept, since failure to provide an origin narrative left the source of the event, and its distinction from other infectious disease events, ambiguous.

The WHO did not offer an explanation of the origins of H1N1 in representing the phenomenon to outside actors. This left the emerging concept somewhat ill-defined, and contributed to the failure of the WHO to successfully enrol other actors into the pandemic. From the WHO's institutional perspective, the question of origin was not considered to be of primary importance, as a future-orientated perspective was adopted in the management of the pandemic. However, in representing disease to outside actors and to the public, a conception of the source of illness is fundamental to the ultimate recognition and understanding of a threat (Herzlich & Pierret, 1987). In the case of H1N1, it was clear from the questions of the press that in the wider sense, origins were considered to be a matter of public importance. This difference in emphasis between the WHO and other actors was a source of contention, as was evident in the section of press conferences where the WHO's representatives fielded media questions.

The WHO's responses to the question of origins were dismissive. For example, in response to one question, it was stated by the WHO's representative:

> I know that there is a fair amount of speculation about where the virus may have originated. I think that right now it is not possible to really know where this virus originated. Most of the virus that we see out there are very similar to each other suggesting that they were a newly emerged virus rather than one that has been around for a while, and has many different variants, but I think it is too early now to speculate about the origins.
>
> (Fukuda, 28/04/09)

And on a separate occasion in response to another query, the lack of emphasis on origins was made clear:

> everyone is always interested about with a new disease. But I would say that at this point we have higher priorities... the kind of investigations that are really critical right now to answer the most urgent issues is how this is evolving, where it is going, what is the impact on people, what steps might be taken to protect people. Nonetheless, I believe that at some point we should come back and try to understand what are the origins of this virus... Very interesting questions but maybe not the highest priority right now.
>
> (Fukuda, 29/04/09)

While the question of index cases and identifying the point of zoonotic transfer were thus perceived to be in the public interest by the media, the WHO tended to disregard these questions in favour of a more pragmatic focus on the impact. Investigating the question of origins was postponed until after the threat had passed.

Nevertheless, some attempts at broadly explaining the process of zoonosis were made by the WHO. For example, it was explicated that

> one of the things to explain is that typically pandemic viruses start from animal viruses in the sense that they become humanized, so animal viruses for whatever reason every once in a while come over from the animal side or some other genes do and then they lead to infections in people and when they progress long enough they really become more human viruses. *And so in a sense that is what a pandemic is.*
>
> (Fukuda, 26/04/09; emphasis added)

In this example, zoonosis itself was set to characterize and distinguish a pandemic event. However, simultaneously the WHO's narrative dismisses these origin narratives as unimportant. Again, in relation to zoonotic disease transfer,

> I think this is [a] phenomenon we have all been observing over that last number of years. If we look at major threats to international public health security over the last three decades, many have emerged from animal origin and diseases which breach the species barrier and establish themselves in humans... The animal/human interface needs to be watched carefully and needs to be managed through the proper risk management and collaboration...
>
> (Ryan, 02/05/09)

Thus, while specific questions of index cases were not investigated (at least, during the pandemic event) by the WHO, the general principle of zoonosis was emphasized as a means of indicating that new reservoirs of disease are ever-present through the existence of novel strains in animal populations. However, the WHO's explanations failed to provide a specific explanation of the case of H1N1 and therefore failed to satisfy the social need to understand causes of disease.

The lack of definition of origins resulted in action towards H1N1 that was unintended by the network-building actor, the WHO. For example, one of the consequences of the lack of clarity surrounding the origin narrative was the actions taken in regards to pig populations. One of the ways in which infectious disease is explained is through reference to the dirt/contagion carried by animals as a part of the discursive process which forms a distinction between non-human animals and humans (Douglas, 1973; Haraway, 1991). As the common name 'swine flu' suggests, the H1N1 strain originated broadly from a porcine host reservoir (known from antigenic typing). This (unspecific) knowledge of the origin of disease resulted in various national governments taking action with regard to the management of pigs, and led to significant confusion over the role of pigs. This again shows the importance of origin narratives in the interaction of actors with a perceived disease threat. In response to such actions, the WHO's representatives needed to emphasize that,

> traditionally, as I mentioned these viruses have circulated in pigs but so far we have no evidence which suggests that these people were exposed to any sick pigs so we don't have any direct connection to swine right now.
>
> (Fukuda, 26/04/09)

Furthermore, specific instances of anxiety surrounding pigs had to be addressed by the representatives.

The desire to understand origins is important. When origins are understood, the source of disease (and therefore blame) can also be distinguished to provide a socially robust account (Nelkin & Gilman, 1991). The lack of a clear origin narrative in the WHO's problematization of H1N1 resulted in confusion and misunderstandings regarding the source of disease. This confusion could be seen in the actions of other elements in the emerging actor network with regard to porcine populations. This is evidenced, for example, in the question of reporter Jamil Chadai of Sao Paulo:

> Could you explain why it is that WHO insists that embargos on pork meat is not recommended and actually that trade can go on, and eating can go on, if you just said that you are actually studying the effects? How is it that you are 100 percent sure that it is not going to be a problem?
>
> (Fukuda, 03/05/09)

In response it was stated by the WHO's spokesperson:

> we know that they [influenza viruses] are not very resistant to heat, meaning as soon as you cook a product that may contain these viruses, they will get inactivated.
>
> (Fukuda, 03/05/09)

In another example, a reporter questioned whether the WHO had not clearly articulated its message on the safety of farming and consuming meat:

> I think you suggest this issue of pigs. Today the parliament in Egypt ordered to slaughter all the pigs in the country. I am wondering if there is not perhaps too much misinformation still out there, that WHO isn't doing enough to combat...
>
> (Fukuda, 29/04/09)

The question of ineffective communication (and, at its root, the failure to articulate a coherent origin narrative) was not addressed. Instead, the safety exposure to pigs was again stated:

> At this point, I want to make it very clear that we do not believe that the infections occurring in people are associated with getting infected from exposure to pigs. This is a different situation from what we saw with avian influenza – the bird flu – in which people got clearly infected by birds.
>
> (Fukuda, 29/04/09)

In this way it is clear that understandings of the previous pandemic threat (avian influenza) and the mode of transmission through bird handling (which was incorrect) resonated within perceptions of swine flu. This explanation from past experience was utilized analogously in lieu of the WHO producing an effective explanation of the origin of H1N1. As a result of this, it was necessary for the spokespeople to reiterate the safety of interacting with and consuming swine throughout the conferences. For example,

> we are dealing with a situation where the people who are getting infected are not getting infected from pigs. Having said that, of course, we always want to be careful and make sure that there is no risk so this is something that we would continue to look at as to whether pigs may pose a risk to some people but this is really not how people are getting infected and this is really important to understand and be clear on.
>
> (Fukuda, 04/05/09)

Even here, there was some ambiguity in the WHO's position. While at the start of the quote it is 'clear', the second half suggests some apprehension. This position was consolidated at some point so that in a later conference the narrative was plainer:

> we have tried to make it very clear that we see this current situation as reflecting transmission of infections from person-to-person. Eating pork is not a danger in terms of getting this infection. I just want to reemphasize this point again...
> (Fukuda, 07/05/09)

The need for the WHO's representatives to continuously make statements to this effect indicates the confusion regarding zoonosis in the case of H1N1. The failure of the WHO to effectively translate an origin narrative into the concept of H1N1 affected the way in which the disease was related to as a whole. It demonstrates a failure to successfully define and distinguish the concept, and a deficiency in the effort of translation. The initial designation of the threat with the name 'swine flu' partially led to the confusion. The effort at naming was again an unsuccessful aspect of the problematization of the phenomenon.

Naming

In addition to the confusion surrounding the role of pigs in the transmission of H1N1, the early adoption of the name 'swine flu' to refer to the H1N1 threat was a barrier to the overall translation of the concept. Naming is an important component of disease construction. The naming of a disease is often pivotal in the understanding of what that disease constitutes (Aronowitz, 1991; Brown, 1995; Karkazis & Feder, 2008). This is seen, for example, in instances where changing the name of a disease changed the social relations surrounding it, or where different interest groups adopt strong stances in relation to a specific label (Beard, 2004; Kushner et al., 2004). A name can be vital to characterizing a new object. Changing names can change the object, and consistent naming aids in maintaining the stability of (perception and interaction surrounding) the entity. As such, naming is an important part of the problematization of a new disease.

The WHO recognized that the naming of diseases can often be value-laden. The organization thus acknowledged that naming presented a difficulty in the case of H1N1. In this respect it was stated that

> since the emergence of the pandemic, the name of the virus has been a difficult issue for many reasons. In the past, we have seen how the naming of viruses by location can stigmatize those locations and we

have also seen in this and in other episodes where associating the virus with one animal species or another, can really cause both anxiety and then fears about food and in this particular instance, about pork.

(Fukuda, 07/06/09)

However, agreeing upon a suitable name had proved difficult, and it was not definitively established until just prior to the declaration of the pandemic. Thus, at the start of the threat, concerns were raised about the name, such as indicated in this press question:

> I just wanted to clarify something regarding names. I have seen that the ECDC [European Centre for Disease Control] is now renaming, saying that they prefer to rename this novel influenza. Does the WHO have an opinion on the name as I can imagine the importance to have a single name that everyone uses in such a situation: if you could clarify the status of the name?
>
> (Fukuda, 28/04/09)

The WHO's representative responded:

> Now in terms of the name, again I think that the naming of new diseases, the naming of epidemics can be very confusing. At WHO we have not initiated any plans to try to change the name 'swine influenza'. This epidemic started basically with that name and the virus that is identified is a swine influenza virus, and we are hopeful that the introduction of new names does not cause any undue confusion. But right now we do not have any plans to try and introduce any new names for this disease.
>
> (Fukuda, 28/04/09)

The following is another instance of the WHO's representative attempting to clarify the problem of naming:

> We know that the situation has been confusing. For example, right now, we know that there are H1N1 viruses which have been circulating in people for a number of years. This is a new H1N1 virus. And we also know that there has been H1N1 viruses which had been circulating in swine or pigs for many years. And this has really led to a complicated situation of what you call a new virus. One of the primary concerns and one of the difficulties of naming such a new

virus is to avoid adverse effects, or stigma associated with a virus name...

(Fukuda, 09/06/09)

As suggested by this quote, one of the problems of naming is that a name is often associated with blame. The WHO's reluctance to effectively name H1N1 at the start of the pandemic was in part a consequence of the importance of naming.

In the end, the more apparently scientifically neutral name of 2009 H1N1 was adopted, and just prior to declaration it was suggested that

what the experts decided – calling this pandemic H1N1/09 virus – was a good way to distinguish it from the current H1N1 viruses and to do so, in a way which was scientifically sound, but also would avoid some of the stigma associated with other options.

(Fukuda, 07/06/09)

Thus the late adoption of a new name did not aid the translation process because it left the phenomenon somewhat ill-defined until after pivotal outside actors had begun to interact with it. That is, interessment and enrolment occurred before stable naming (which is part of successful problematization, the first stage in translation). The lack of decisiveness with regard to the naming of H1N1 was one element of the unclear problematization as a whole. The WHO's actions directly led to the failure to stabilize H1N1, which left the concept contested and undermined the WHO's attempts to enrol other actors.

There were several aspects of the WHO's construction of H1N1 which lent to its ineffectual translation. A consensus upon an appropriate name was not reached early on in the process, which increased the instability of the concept as a whole, given the importance of naming in building consensus around a new actor network. Additionally, as shown above, the WHO failed to address the problem of origins, and thereby did not clearly situate and explain the disease. Furthermore, while historical comparisons were mobilized relatively effectively (though were not altogether uncontested or unambiguous), the WHO failed to clearly distinguish H1N1 from seasonal influenza. These factors all contributed to the eventual vulnerability of the concept of an H1N1 pandemic, as the combination of these inadequacies resulted in the failure to effectively translate H1N1 as a pandemic threat. To be successfully translated, the concept of H1N1 needed to be definitionally stable. However, as indicated by the definitional problems listed above, the WHO had attempted to mobilize the concept and enrol other actors into association with the concept before this definitional stability was achieved. As will be demonstrated in the subsequent chapters of this book, the concept of the H1N1 pandemic ultimately failed to gain traction and support within the wider global health arena. The inadequacies in the initial translation of the concept resulted in its instability as a scientific fact and, ultimately, its contestation by other actors.

3
Risk and Scientific Uncertainty

As has been demonstrated, there were several aspects of the WHO's attempt to construct a stable actor network for H1N1 which rendered the emerging concept vulnerable to contestation. In Chapter 4 another pivotal feature of the construction of H1N1 will be discussed – the WHO's representation of risk. From the perspective of the WHO, the H1N1 pandemic was of concern because it indicated a formidable global health threat. In order to mobilize this narrative across the actor network, the production of an effective risk discourse was critical. Sociologically speaking, since the conception of 'pandemic' signifies a considerable threat to global health, articulating H1N1 as a 'risk' was a key element in successfully translating the virus into the category of 'pandemic'. However, as this chapter will demonstrate, the WHO failed to mobilize an effective risk discourse. In fact, the risk narrative produced by the WHO was subject to multiple revisions and replete with fundamental inconsistencies.

Theoretically, the failure of the WHO to mount a strong claim regarding the risk of H1N1 is fruitfully explained using co-productionist theory. Co-productionist theory claims that the production of scientific knowledge under contemporary conditions of risk and uncertainty (referred to variously as post-normal science or Mode 2 science) denotes a marked departure from previous practices of science (Funtowicz & Ravetz, 1993; Funtowicz & Ravetz, 1994; Jasanoff, 2004a; Lenhard et al., 2006; Ravetz, 2004). The scientific knowledge that is produced under conditions of risk is itself uncertain and replete with contradictory disciplinary conceptualizations. This is because the (necessarily) limited knowledge surrounding risks, and their fundamentally future orientation, produces diverse explanations of them. Within this climate of scientific uncertainty, the WHO needed to adhere to one of many possible perspectives in constructing its risk discourse. The perspective favoured was a reference to severity as measured through geographic spread. This ultimately proved to be an inaccurate model for portraying the effect of H1N1 upon global health. This lack of a socially robust risk narrative underpinned the instability of the H1N1 actor network as a whole.

The sociology of risk

To translate H1N1 as a legitimate threat, and a 'pandemic', the WHO needed to mobilize an effective risk discourse surrounding the emerging actor network. Here I analyse H1N1 in the context of the scientific/institutional construction of risk. Furthermore, I show how recent developments in the structuring of science and the effect of large-scale societal changes (e.g. globalization) served to construct the problem of H1N1. To explain the WHO's risk discourse surrounding H1N1, this chapter will employ a theoretical framework which is derived in part from the acknowledgement of the proposition of the risk society. This is complemented and extended by co-productionist theory, which is underpinned by constructionist conceptions of risk, conceptualizing risk as both constructed and ubiquitous in contemporary society. Co-productionism therefore builds upon macrological constructionism (e.g. Beck, 1992; Giddens, 1991) but is also concerned with the micrological procedural details of the construction of 'things' (as adopted from ANT). Co-productionist theory also advances further on traditional risk society theories in that it helps to demonstrate the way in which the structure of scientific knowledge serves to produce individual instances/constructions of risk.

Co-productionist theory demarcates its field of interest as concerning the production of scientific knowledge within the wider societal context. The primary argument of co-productionism is that there is a reflexive relationship between scientific knowledge production and society at large, such that each serves to create the other (Jasanoff, 2004a; 2004b). In interrogating risk, co-productionists adopt constructivist theories of risk, agreeing that contemporary society is fundamentally risk-laden, and accept that risk has become magnified in contemporary society as a starting point for their explanations. Acknowledging this point, co-productionist theorists serve to illustrate the impact of risk upon the practice of scientific knowledge-making. Further, the pervasiveness of risk within contemporary society underpins the production of scientific knowledge.

The study of uncertainty is now central to the task of scientific research. Whereas science was previously concerned with advancing certainty/control over the natural world, contemporary science is defined by the concept of uncertainty (Funtowicz & Ravetz, 1993). This is reflected in what Latour (1998) refers to as a fundamental shift from 'science' towards 'research'. According to this argument, scientific endeavour was previously focused upon producing 'science'. This means that the knowledge which was produced attempted to answer the questions born of science, which were tied to the enclosed research objectives of scientific disciplines. In short, scientists were interested in answering the questions of how the world worked as according to the ontology and epistemology of their disciplinary paradigms. The questions of 'science', while necessarily indicative of wider societal norms and values (i.e. concerning what constitutes epistemologically privileged knowledge), were thus products of the scientific community and particular scientific disciplines. The contemporary emphasis on 'research' over 'science', in part a reflection of the societal concern with risk, represents a fundamentally different manner of scientific knowledge-making. Following Latour, 'research' consists of scientific investigation aimed at addressing the needs/demands of society, including (and especially) risks. The questions of 'research' are therefore born out of the wider society, not of any specific scientific community.

The problems of 'research', being real-world problems, encompass a wider range of uncontrollable variables than the problems of 'science' (which often investigate highly specific and controlled phenomenon). In this way, as will be discussed in more depth later, the questions of 'research' also tend to transgress disciplinary boundaries because they are not bound to the concerns of a single discipline (Lynch, 2004; Nowotny, 2003b; Shackley & Wynne, 1996; von Schomberg, 1993a). Latour's conception of the science/research distinction is a compelling theoretical argument, and it serves to explain the apparent rise of problem-centred intellectual investigation over basic research. The argument refers to overall shifts in the funding mechanisms of research institutions as instigating this change (Latour, 1998). Co-productionist theory, however, emphasizes the effect of risk upon the prevalence of 'research' (Jasanoff, 2004b; Nowotny et al., 2001).

From a co-productionist perspective, risk is an important feature of contemporary scientific work. The structures of science are responsible for identifying and managing risks. The societal insistence that science works to alleviate risks therefore has fundamental consequences for the structuring of science, and it reflects a movement from 'science' towards 'research'. It is debateable whether the co-productionist emphasis upon risk in scientific structures is entirely valid. Although it is true that risk is central to contemporary scientific investigation, it could be reasonably argued that this is not a new phenomenon. Especially when considering the field of health research, as is relevant to the present case of H1N1, the investigation of risk has always been embedded into medical science. Partially, this is because medical research from its genesis has studied people as its subjects, and, as such, risk has always been embedded in its work (as opposed to disciplines such as physics, where proof of a shift towards 'research' from 'science' might indeed provide evidence of a wider change in scientific structures). Furthermore, the questions of investigation within medical science have always been the questions (and norms and values) of society, and the 'co-production' of medical science and society has therefore always occurred.

However, despite these critiques, co-productionist explanations are useful for the present analysis simply because H1N1, as a medical 'problem', is clearly and directly embedded within risk discourses and the science of risk (in this case epidemiology). Despite some reservations regarding the extension of co-production over other fields of scientific knowledge production, when considering the case of H1N1, the perspective carries explanatory weight. Co-productionist theory is especially useful in its relationship with the idea of risk. Early constructionist explanations, especially Beck's risk society, tend to include some conceptual ambiguity and contradiction in attempting to explain the nature of risk. For example, Beck argues heavily towards a constructionist conceptualization of risk, but simultaneously suggests that 'natural threats' (which would include infectious diseases such as H1N1) somehow carry a different ontological status from that of constructed 'risks' (Beck, 1992; Mythen, 2007). As is argued in this book, even 'natural' threats are socially constructed and perceived, and are no more or less objectively 'real' than the risk of pollution or environmental waste. Co-productionist theories circumvent this theoretical dilemma altogether. What is important here is not whether the risk is 'real' or not. Rather, the attitude of uncertainty which surrounds risk is what underpins scientific investigation. Whether the risk is 'real' or 'constructed' is beside the point; within co-productionism, what is under discussion is the scientific attempt to produce knowledge under conditions of uncertainty (which occurs equally whether the threat is 'real' or perceived). The societal perception of risk (again echoing the risk society thesis) produces a call for 'research' surrounding the risk, which structures the way in which the scientific investigation is conducted.

Furthermore, since the investigation of risk must transgress disciplinary boundaries (given the multiplicity of variables inherent in real-world and uncertain events), the contemporary study and resolution of risk requires a great degree of interdisciplinarity (Nowotny, 2003a; Saloranta, 2001). However, the existing structures of science tend to be discipline-bound, rendering risks (such as H1N1) as problematic subjects of investigation. Although risk-managing organizations, such as was the case with the WHO, attempt to form expert panels which range across multiple required disciplines, often one perspective gains ascendency over the others (Shackley & Wynne, 1996). This is unsurprising, given that the primary assumptions (and understandings of what constitutes important data and information) within one discipline will vary from another, even within disciplines which may all be described broadly as medical science (Lenhard et al., 2006; von Schomberg, 1993a). For example, the concerns, interests and assumptions of the virologist would be vastly different from those of the epidemiologist or the public health specialist. Accordingly, as suggested by co-productionist theory, organizations which manage risk need to choose between these competing perspectives in defining and articulating the threat in a socially robust manner. In the case of H1N1, the WHO chose to privilege epidemiological conceptualizations of 'severity' in defining the risk of the pandemic. 'Severity' came to be presented by the organization as synonymous with risk/threat, and was defined through the criterion of geographical spread. The language of geography and severity was thereby embedded in the WHO's construction of the risk of H1N1.

Severity/mildness

Severe disease is understood as a significant risk (and conversely, a mild disease is associated with an insignificant risk). When assessing the potential risk posed by a pandemic threat, the epidemiological terminology of severity was key to the WHO's discourse. In the case of H1N1, the concept of mildness and severity was a pivotal point of discussion. In fact, in addition to the controversy surrounding the use of vaccines (see Chapter 5), competing assertions of 'severity' versus 'mildness' was a primary site of contestation among global actors. Importantly, the WHO's conceptualization of severity was inconsistent and was subject to drastic changes through the course of the pandemic. This is because 'severity' was a black-boxed concept that came to be opened through the contestation of H1N1. This resulted in a fundamental shift from the initial naive/unproblematic use of the term (when the concept was still black-boxed) to a series of redefinitions which rendered the concept increasingly complex (when the concept became contested). Finally, the organization attempted to ultimately resolve the problem of severity by abandoning the concept altogether. However, this also proved to be an untenable strategy, given the strong linking of severity to risk within explanations of infectious disease.[1]

In the earlier texts it was evident that the WHO's representatives referred to severity in a straightforward and unproblematic manner, and regarded it as a defining feature of the risk posed by a pandemic. For example, it was asserted that

> If we ask ourselves what are the main questions about the disease that we would like to know about, I think the most important one is how often does it lead to severe disease.
>
> (Ben Embarek, 04/05/09)

The understanding of severity was relatively unsophisticated, and corresponded to the typical lay usage of the term, which relates severity to serious clinical manifestations. Thus, at the early stages of the threat, the WHO proposed that 'The severity of pandemic has to do with when people get infected, how often are they going to develop really severe disease' (Ben Embarek, 04/05/09) and asserted that

> One of the things that we are trying to do is to identify what are the most pressing scientific issues, and then try to address them as quickly as we can. I think that right now, the severity of illness – the clinical features of illnesses – is one of the most important questions...
>
> (Ben Embarek, 04/05/09)

In this way the concept of severity was taken for granted in early conceptualizations of pandemic risk. It was also clear that severity was regarded as fundamental to understanding the threat.

Simultaneously, in the early texts it was evident that the WHO placed an emphasis on the importance of determining severity. This is demonstrated in the following quote, which stated that

> The other question that has come to WHO is: 'Is severity important?' Of course severity is important. The whole reason why we take action against diseases is because they harm people. If diseases are relatively mild, like colds, then we take certain kinds of precautions, if diseases are very severe, such as avian influenza or HIV, then we take

another level of precautions. Clearly severity is an important concept for public health and how we deal with these issues.

(Fukuda, 26/05/09)

Thus, throughout the early usage of the term by the WHO, 'severity' was suggested to be a fundamental concept which signified and defined the risk that was posed to a population by a pandemic. 'Severity' was black-boxed – taken to be understood but never fully articulated.

However, as the H1N1 pandemic unfolded, this usage of the term became unsustainable. In the early usage it was clear that the interest in H1N1 as a threat stemmed from its probable ability to produce severe disease – the potential severity characterized H1N1 as a risk. However, the pandemic failed to manifest as 'severe' in this black-boxed sense of the term. Therefore, as a reaction to the unfolding of events, the linking of severity and risk was increasingly disassociated. This occurred through a redefinition both of the term 'severity' and of the (previously) implied correlation between severity and risk.

The prior link between severity and risk was unsustainable once it was recognized that H1N1 was mild in its effects. This was because if the disease was mild (not a risk) then, by extension, there was no justified point of concern – the risk would be eradicated through a lack of severity. The WHO, after announcing a pandemic state, therefore needed to maintain concern (i.e. sustain a risk discourse) regarding H1N1 while simultaneously encompassing mildness. This occurred through a redefinition of 'severity'. One of the ways in which this occurred was through reference to other concepts in epidemiology. As previously discussed, one of the characteristics which was used to identify a pandemic threat was the distinction between pandemic and seasonal influenza. In the absence of widespread severe disease, this distinction was relied upon heavily by the WHO to sustain a high risk alert. It was therefore suggested by the organization that one reason why H1N1 should not have been referred to as mild was because it was different from seasonal influenza. Thus it was stated:

I want to point out that we are not dealing with seasonal influenza. I think there is a lot of confusion and a lot of comparisons made but this is one of the basic points. There are some features with this pandemic that we are seeing which are similar to seasonal influenza [but, after elaborating etiological and epidemiological differences that]...this is a very different pattern than we normally see with seasonal influenza.

(Fukuda, 05/11/09)

This difference in epidemiological patterning served as a basis for suggesting that H1N1 was not mild. It was also an attempt to prevent the loss of the concepts of 'risk' and 'severity', which were key components of the WHO's reaction to H1N1. Risk and severity were redefined to suit the unfolding circumstances.

The WHO also highlighted several other suggested 'pandemic-like' aspects of the epidemiological pattern of H1N1 to accentuate the risk of that disease. Notably, the unusual incidence of illness in the young and otherwise healthy was emphasized. For example,

> One thing I think which is true is that so far, among the cases being seen everywhere, including the countries with the large number of cases, is that the people being infected continue to be relatively young people.
>
> (Fukuda, 05/05/09)

The presence of H1N1 infection in the young was presented as a key distinction. In the case of H1N1, 'It is probably fair to say that approximately half the people who have died from this H1N1 infection have been previously healthy people' (Fukuda, 09/06/09), which represented a key way in which the WHO began to signify the threat of the pandemic:

> The reason why we mentioned this point is that because usually healthy adults, when they get flu – I mean seasonal flu – they usually have mild symptoms. It is very rare that young adults become severely sick with flu. This is a new feature with this virus. But the numbers are quite limited so far.
>
> (Briand, 08/05/09)

As these quotes illustrate, the risk surrounding H1N1 was characterized as existing not in its clinical severity but in its unusual incidence in the demographics of young, healthy adults. This was emphasized even given the fact that 'the numbers are quite limited' because it provided an important point through which the WHO could distinguish the pandemic.

This emphasis on epidemiological incidence was mobilized to guard against criticisms by the media and the (perceived/imagined) public reaction to the WHO's characterization that the risk and severity of the pandemic were inaccurate. Thus, while it had to be acknowledged that most cases of H1N1 manifest in minor disease, it was asserted that this fact that most people recover from illness has led some people to speculate that this is really a very mild situation and really do dismiss the pandemic infection... at WHO we remain quite concerned about the patterns that we are seeing, particularly again, because a sizable number of people do develop serious complications and death and again we are seeing most of these occur in people who are younger than 65 years – a picture which is different from seasonal influenza.

(Fukuda, 05/011/09)

Again the 'mildness' of the disease was evident but the impact, and its distinction from seasonal influenza, was emphasized. Though many features of H1N1 mirrored seasonal flu, those which distinguished the pandemic state were highlighted:

we understand that this disease is mild in the majority of cases, however, we will have some serious cases, mostly in people with underlying conditions, which is close to the pattern we see in seasonal influenza, but we can expect also some cases in people, previously healthy, who will suffer from this virus directly.

(Briand, 08/05/09)

While it had been suggested (using initial black-boxed definitions) that risk equated to severity of impact, as the pandemic unfolded, other epidemiological distinctions between pandemic and seasonal influenza strains were emphasized as an important marker of risk.

Another result of these attempted distinctions was that even a few cases of severe clinical manifestations of H1N1 were proposed as justification of the WHO's strong risk discourse surrounding the H1N1 virus. Thus

there are some exceptions that must be the focus of particular concern. For reasons that are poorly understood, some deaths are occurring in perfectly healthy young people. Moreover, some patients experience a very rapid clinical deterioration, leading to severe, life-threatening viral pneumonia that requires medical ventilation.

(Chan, 11/08/09)

As in the statement above, the initial association of risk with severe disease became reconceptualized in increasingly complex ways to include propositions of the importance of other epidemiological distinctions. As has been demonstrated, one subnarrative of the WHO's account of risk was that H1N1 represented a threat simply by being different and novel (i.e. through its distinction from seasonal influenza).

The second key way in which a risk discourse was sustained by the WHO was to directly redefine severity. The concept of severity was unravelled and rendered increasingly complex as the event wore on. This occurred through the WHO's assertions that while severity was indeed important, it was also difficult to measure objectively. In addition to its distinction from seasonal influenza, this was a key narrative, where it was suggested that severity is an unstable and inconsistent characteristic of a pandemic. In these instances, while the importance of severity continued to be emphasized, the difficulty in its assessment was also highlighted. In this way it was suggested that

> at the start of a pandemic, one of the things that we try to get is an assessment of the severity but it is important again to remember that the properties of 'flu viruses can change over time. They can go from mild to being more severe as time goes on and they can also move from being more severe to less severe over time. It is way too early right now to predict whether we might see a mild pandemic or a severe pandemic, but again we will keep you updated...
> (Fukuda, 26/04/09)

Thus 'severity' was still equated by the WHO here with the clinical manifestation of the illness, but was also expressed as a changing (and thereby uncertain/risk-laden) characteristic.

This notion of changing and evolving severity is consistent with the WHO's attempted characterization of H1N1 within strong discourses of risk, and it was utilized as a justification for the concern regarding the virus. Thus, in response to early criticisms of mildness, where critics pointed to the successful resolution of the vast majority of initial cases, it was suggested by the WHO that

> In terms of the mildness of the cases out there and whether people may take a pandemic seriously or not seriously, I think the main point I want to make here, the most important point to make here, is that it is probably premature to think of this as a mild pandemic or as a severe pandemics, and it is very clear that we cannot predict what the cause of this [severity/mildness] will be.
> (Fukuda, 29/04/09)

In this way, while to some extent ill-defined and black-boxed conceptions of severity were still upheld, it was suggested that severity is a characteristic that is indeterminate, being difficult to measure and liable to change.

Another way in which the WHO attempted to maintain its risk discourse was through suggestions that risk can be ameliorated/minimized through human intervention. These claims were made after it had become clear that the virus was not going to produce a severe disease outbreak. This tactic represents a departure from those previously described because it begins to incorporate the sense that H1N1 was in fact a mild disease. In this way it was asserted that the H1N1 pandemic did indeed unfold as a mild event, but that this was only due to the impact of the efforts to minimize risk. It was asserted that

> Now that we are a few weeks into it, the picture is clearer. We have seen that in fact most people so far have developed a milder form of the illness, but again we have noted that deaths are there and the numbers have increased somewhat.
> (Fukuda, 11/05/09)

More tellingly, this minimized harm was suggested to be the result of the organization's own actions:

> This is really an important point: it is often hard to see what it is that you have prevented by doing so much work, but it is exactly for this kind of situation. So if things turn out so that few people die, this would be the best of all possible outcomes. But in the meantime we will continue to work as hard as we can to make sure that countries are as prepared as possible. It is not just a state of whether you are prepared or not, it is something that you continuously work at...
> (Fukuda, 11/05/09)

Thus it was suggested that H1N1 had indeed posed the threat of a severe disease but that the preventative reactions of the WHO and its member states minimized the impact of the pandemic. This argumentative strategy thereby served both to highlight the continuing risk of pandemic threats and to reinforce the necessity of the WHO's work. However, it was not among the most emphasized explanations offered by the WHO, probably because the disease was often also mild in countries where the organization's advice had not (at least significantly) been implemented.

More prominently, throughout these events, the largely unproblematic/black-boxed use of the term 'severity' in earlier texts was more completely replaced through an increasingly complex series of attempts at definition and redefinition. Severity changed from a constant indicator of serious disease towards a characteristic that in itself was uncertain. In this way the WHO texts highlighted that

> The question that is really on people's minds is what can we say about the severity of the illness at this point. I think that the information to date clearly points out that this infection can result in anything from a very mild illness, where people do not need to be hospitalized and generally recover without any complication after several days, to fatal illness.
>
> (Fukuda, 29/04/09)

Even here, early on, the WHO started to introduce ambiguities in the discourse surrounding severity, in this case highlighting its case-specific nature.

Furthermore, the inconsistent nature of 'severity' was also presented through the suggestion that it varies between different geographic regions. This argumentative strategy suggested that severity was not a simple characteristic that could be easily recognized since

> it is also clear that what is severe is one country is not necessarily severe in another country. This is one of the lessons that we have learned from many outbreaks certainly one of the lessons from influenza.
>
> (Fukuda, 26/05/09)

These nuanced additions to the concept of severity indicate the fragility of the term across the course of the pandemic. Due to the fact that severity was interlinked with (significant) risk, the WHO faced difficulty in abandoning the concept altogether. Nonetheless, the failure of H1N1 to manifest itself through clinically severe disease in the vast majority of cases resulted in the WHO being forced to redefine the meaning of 'severity' in order to attempt to maintain a persuasive risk discourse. 'Severity' was then described as being spread across a spectrum, rather than being a fixed characteristic, with H1N1 being 'moderately severe' (note that this is still not 'mild') in most cases.

In this way the WHO started to acknowledge that H1N1 did not represent a severe disease. Over time the acknowledgement of mildness was co-opted into the concept of a spectrum of severity. Thus, again reinforcing suggestions of the impermanence of severity, it was noted that

On present evidence, the overwhelming majority of patients experience mild symptoms and make a rapid and full recovery, often in the absence of any form of medical treatment.

(Chan, 17/06/09)

However, it was simultaneously suggested that severity has historically been variable through the course of a pandemic:

> One of the lessons that history has shown us is that pandemics span from being mild to being extremely severe. My own sense right now is that it is probably too early to make a pronouncement about what kind of pandemic we may see. It is entirely possible as the ECDC has commented we might see a very mild pandemic, and that would be the best of all possible situations short of this current situation simply stopping and disappearing, but I do want to provide a cautionary note. The worst pandemics on the 20th century occurred close to the beginning of that century and it also started out as a relatively mild pandemic, with illness that was not very much noticed in most cases and then became a very severe pandemic and one of the most severe infectious diseases the world has recorded.
>
> (Fukuda, 28/04/09)

This quote reflects an attempt to strengthen the claim of potential severity, despite the fact that other disease-managing institutions (here the ECDC) had started to produce narratives of a mild disease. More starkly from previous descriptions, it was eventually conceded within this subnarrative that the H1N1 pandemic was indeed mild in nature:

> But I think the other point is simply true that it is quite possible to have a pandemic on the milder side and if we are experiencing that, and if the number of serious cases is kept down, and this is something, again, something for which we should be thankful.
>
> (Fukuda, 03/12/09)

However, this mildness did not mean that H1N1 did not constitute a pandemic:

> we really see that this pandemic is on the less severe end of severity. Again we don't really know what the final impact is and we won't know what it is until a year or two after the pandemic is over but it appears to be on the less severe side of the spectrum of pandemics that we have seen in the 20th century.
>
> (Fukuda, 14/02/10)

This demonstrates that the WHO's attempts to link H1N1 to severity were ineffective given the actual nature of the disease. As a result, the WHO attempted to mobilize a depiction of pandemic influenza which suggested that such events are not necessarily severe. Pandemics, here, are presented as lying on a spectrum of severity. Such statements served to mobilize the argument that preparatory actions remain of vital importance (by pointing out the variability of severity) while co-opting a growing account that the impact of the H1N1 virus may not be great.

It is clear therefore that the WHO faced great difficulty in mobilizing an effective discourse of severity. The H1N1 pandemic did not possess the common attribute of serious clinical disease, making it difficult to characterize the pandemic within discourses of risk. As a result, by the final stages of the pandemic, attempts were actually made by the WHO to abandon the concept of severity altogether by suggesting that it is an ambiguous and meaningless term. This attempt at abandonment is highlighted in Harvey's speech as the chairperson of the International Health Regulation Review Committee:

> When you think about severity, you have at least three problems... The first is defining what you mean by severity. So, are you talking about mortality? Are you talking about morbidity, or illness? Are you talking about some combination of the numbers and the severity in meaning 'severity'? What do you really mean by it?...
>
> Second challenge is, how do you measure it? Not simply in theory – but how do you measure it practically and in real time, in a way that can be used to inform your decisions?
>
> And third, how do you account for the variety in severity in different settings at the same time?
>
> (Fineberg, 14/04/10)

This passage demonstrates the WHO's attempt to completely problematize the concept of 'severity', and in doing so to delink the association between severity and the risk posed by infectious disease threats. In this way, towards the end of the pandemic, there were clear attempts by the WHO to distance itself from early taken-for-granted conceptions of severity and even to abandon the concept altogether. Severity was thus radically reconceptualized from a black-boxed characteristic which was clear and investigable to a contested characteristic which was complex and impossible to measure.

The abandonment of 'severity' is also demonstrated in the WHO's attempts through the later stages of the pandemic to suggest that pandemic events are in fact often mild – that is, that severity is not an important or defining characteristic of a pandemic. This suggestion that severity does not determine whether a disease constitutes a pandemic was reinforced through references to historical events. Here it was suggested that 'we have also noted that when we look at the Twentieth Century the experience has been that pandemics can range from the relatively mild side, to being on the more extreme side' (Fukuda, 03/12/09). Thus a pandemic was depicted as an event that is not necessarily of high clinical impact.

In this way, towards the later stages of the pandemic, the WHO suggested that severity was in fact not an important conceptual feature of pandemics, and the concept was abandoned. The consequences of this attempt to abandon 'severity' will be elaborated on further in the discussion of the controversy surrounding the pandemic declaration and pandemic phase definitions. However, it is argued here that the strong assumed correlation between severity and risk resulted in the inability of the WHO to abandon the concept of severity and simultaneously retain a strong risk discourse. The WHO attempted to maintain and negotiate the linking of risk and severity through its distinction from seasonal influenza, an emphasis on the risk posed to certain populations, and an attempted redefinition of severity as a contiguous characteristic. When this became unsustainable, the WHO eventually attempted to discard the notion of severity altogether. However, this proved untenable and bolstered the criticisms of the WHO's account. After all, if the pandemic did not represent a severe risk, then it is difficult to sustain the argument that large-scale risk minimization measures were justified. Thus the organization's failure to mobilize the concept of 'severity' in a consistent and robust manner partially underpinned the ultimate fragility of the concept of H1N1 itself and indicates an institutional failure to effectively mobilize the pandemic within the discourses of risk. Partially this institutional failure was a result of the context of scientific uncertainty in which the WHO was making its early management decisions.

Uncertainty and the need for (scientific) information

To explain why the WHO failed to produce an effective risk discourse surrounding H1N1, the context of uncertainty in which decisions were made is of vital importance. Due to the novelty of the H1N1 virus, scientific evidence surrounding it was scant and emerging, especially in the early stages of the events. This scientific uncertainty in part framed the WHO's reactions. In the absence of the capacity to mobilize the idea of objective scientific 'facts' in framing the pandemic, the WHO instead created a risk discourse that emphasized the threat of uncertainty.

Research produced under conditions of risk is always in itself uncertain due to the sheer number of variables and contingencies involved in the (potential/perceived) manifestation of a risk (Miller, 2004; Nowotny, 2003b; Shrader-Frechette, 1993). Risky phenomena such as H1N1 are novel, complex variable and, by definition, ill understood. However, the structures of science, which are endowed with the task of explaining such phenomena, are ill equipped to research risk. Although it is assumed that science can provide the answers, scientific investigation in practice cannot produce solid and complete evidence surrounding risk. At best probabilistic models are achieved, which are essentially untestable and tentative (Funtowicz & Ravetz, 1993). As such, the scientific knowledge produced surrounding a risk tends to be anecdotal (empirically – and here especially epidemiologically – based upon limited initial evidence) and speculative (theoretically – based on hypothetical models of the future due to a lack of empirical data). However, once risk had entered the social consciousness (i.e. after Mexico reported its first cases), it necessitated action. In addressing the risk of H1N1, an appeal to scientific facts became politically necessary for the WHO in order to be perceived as appropriately managing the case, and to provide evidence for decisions and actions. Simultaneously, the collection of objective empirical data was in fact impossible because there were too many contingent variables (and competing scientific perspectives) involved (Nowotny, 2003b; Nowotny et al., 2001).

It is clear that in order to justify the attention placed upon H1N1, the WHO needed to emphasize the threatening nature of the pandemic. Consequently, in addition to the (unsuccessful) narrative surrounding severity, more explicit references to risk and uncertainty were adopted by the WHO as a strategy to locate H1N1 as a genuine threat. The virus was constructed as a matter of concern because of the uncertainty surrounding it. Thus the notion that 'this is a time of great uncertainty' (Chan, 15/05/09) was emphasized by the WHO's director-general. The term 'uncertainty' itself appeared often in Chan's speeches, and in other WHO texts, both in terms of the future course of the pandemic and in terms of a lack of scientific data surrounding such events (for a few examples, see Chan 04/05/09; 18/05/09; 11/06/09; 17/06/09). It was suggested that

> Whenever we see a new disease, whenever we see any kind of large series of outbreaks caused by viruses that are new to the scene, we are in a period of great uncertainty. This is true of the current period

Risk and Scientific Uncertainty 47

right now. One of the difficulties for decision-makers and countries and public health institutions is that they need to make decisions, they need to move ahead even though many things are not so clear or are not known.

(Fukuda, 09/06/09)

In this way the WHO emphasized the risk of H1N1 by highlighting the uncertainty of the situation. It was also acknowledged that this uncertainty framed its response. The presence of scientific uncertainty was reflected in one of the key management techniques of the WHO as the continuous collection of epidemiological data from affected nations. Widespread monitoring and surveillance were suggested to be one of the ways in which to assuage the unpredictability of the pandemic. Constant collection of data formed an important means through which to minimize the uncertainty of H1N1, and served to justify the WHO's role and actions.

Thus one of the most prominent ways in which the notion of risk was expressed during the early press conferences was through reference to the need for, and lack of, information surrounding the threat. In this way it was stated that 'We know that the need for information is very great and we hope to fill that gap...' (Fukuda, 26/04/09) and that 'What we need most of all, right now, is information... [which] helps us assess and manage risks' (Chan, 18/05/09). Information was viewed as a critical resource in dealing with infectious disease threats. The WHO's suggestions regarding the need for information can be understood in the context of Beck's theorization that a 'risk' is fundamentally characterized by a perceived lack of scientific certainty (Beck, 1992). This is furthered by the co-productionist claim that a consensus upon scientific data is difficult to attain surrounding a risk. Correspondingly, the WHO asserted that the risk and uncertainty of H1N1 must be minimized by the organization through efforts aimed at gathering information surrounding the threat. However, this was a difficult narrative for the WHO to mobilize since that data could not easily be collected.

As a clear example of the emphasis on collecting information, when asked a question about the most effective preparation strategies by a journalist, Fukuda suggested that

when you are facing a new disease threat probably the single most important thing, more than drugs or anything, is just information. If communities and families have information, if countries have information, that is the most powerful thing that you need in the beginning. Without that, you are really in the dark, you do not know what to do, you cannot understand what is going on. I think that you can see, certainly in this response, the effort by a large number of organizations is to get out information as quickly as possible. This is probably to a greater extent than has happened in many outbreak situations in the past.

(Fukuda, 07/05/09)

Information was thus presented by the WHO as key to managing the risk of H1N1. Furthermore, while it was asserted that (through the WHO's surveillance) more information was available during the H1N1 pandemic than had been available for past threats, continuing efforts at gaining information were depicted as necessary to minimize uncertainty.

In particular, the need for further epidemiological information was constantly asserted. It was proposed that in order to effectively combat the threat, a greater level of scientific understanding about H1N1 was essential. In this way it was maintained that an

> area which we are focussing pretty heavily on, is what is the science. And when we are dealing with a new disease we can look at how things develop, we can describe what is going on, but we really want to understand why, because it is the 'why' which is going to give us a handle on how do we handle this better, how do we treat it in a really scientific way, but science does not come overnight.
>
> (Ben Embarek, 04/05/09)

The WHO therefore aspired to react to the situation 'scientifically'. This construction of risk as a matter to be combated by scientific 'fact' was an important motivating factor behind the emphasis upon molecular (viral) and especially epidemiological characteristics. Mirroring this, it was suggested that

> We really need to understand a bit more about epidemiology, we want to understand a bit more about the behaviour of these viruses and we want to understand to the extent that these viruses cause mild infections, and the extent to which these viruses can cause serious infections.
>
> (Fukuda, 26/04/09)

Greater scientific awareness (provided by the WHO) was therefore depicted as key to minimizing the risk of H1N1.

In another example of the WHO's portrayal of the need for scientific evidence, it was suggested that the organization should play an important role in minimizing uncertainty. For example, it was asserted that

> much of what is going on now reflects the fact that so much planning and preparedness has gone on for the past few years. I think that it is still a confusing situation, we cannot come to you and say: 'we understand everything which is going on'. But I do believe that if we had not had all those preparations, if we had not worked so hard to get the information out quickly, if countries had not been thinking about what to do in this kind of situation, in fact we would have had much more confusion and, in many ways, the severity would have been greater.
>
> (Fukuda, 11/05/09)

In this way the WHO proposed that its work in collecting and providing information in itself had led to a reduction in severity and risk. Such a statement could be viewed both in relation to the societal perception of risk (risks are unknowns and providing information about them renders them less threatening) and to the position of the WHO within global health (see Chapter 7). According to the contemporary model of global public health, the primary task of the WHO is simply to collect and convey information. In this way it is unsurprising that the spokespeople emphasized the importance of information in the context of risk minimization. Information and lack of information were therefore proposed both in depicting a justified threat and in depicting the essential role for the WHO in alleviating the threat.

The minimization of risk and the importance of the role of the WHO were therefore emphasized through the value placed upon information. Thus it was stated that

> One of the interesting things about this whole situation is that the amount of information available on what is unfolding is really probably unprecedented. There is more information available about the epidemiology, about the viruses, than has ever been true certainly for a global outbreak like this.
>
> (Fukuda, 14/05/09)

It was also suggested that

> There has probably been an unprecedented amount of information made available about the evolving picture, and the clinical findings,

50 *Pandemics, Science and Policy*

about scientific findings, than any other large scale outbreaks in history.

(Fukuda, 22/05/09)

It was thereby suggested that (due to the work of the WHO) there was an unmatched amount of information available in the case of H1N1. Simultaneously, despite this large amount of information the high degree of uncertainty surrounding the threat did not significantly diminish. As exemplified by the quotes above, it was asserted that the indeterminacy of H1N1 was not due to an inadequacy in collecting and processing information (which would indicate that the WHO was not effectively fulfilling its function) but rather due to the changing nature of the threat. That is, a high degree of scientific certainty surrounding the risk could not be achieved. First, in relation to the unstable nature of the pandemic situation,

> The typical picture for any influenza virus is that it causes both mild 'flu cases and severe 'flu cases and so again the picture is changing and there are very big gaps in what we understand about the clinical impact of this virus. This is one of the things that we will watch very, very carefully and keep you up-to-date on as our understanding develops.
>
> (Fukuda, 26/04/09)

> Clearly there is no standard picture for how these things develop and this is a new influenza virus, so we really don't know how this one will evolve and how the disease related to this one will evolve.
>
> (Fukuda, 27/04/09b)

Second, with respect to the unpredictability of H1N1,

> The virus writes the rules and this one, like all influenza viruses, can change the rules, without rhyme and reason, at any time.
>
> (Chan, 11/06/09)

These quotes suggested that it was the nature of the influenza virus, and therefore the pandemic event, to be unpredictable. As such, information needed to be collected constantly. They also allude to the scientific uncertainty regarding H1N1, which, according to the co-productionist explanation of the science–policy interface, underpinned the WHO's orientation to the problem.

The WHO narrative of uncertainty attempted to accomplish a number of tasks. It served to describe the pandemic as an unclear and threatening situation. Following the description, it was suggested that the collection of scientific information was necessary to minimize the risk – simultaneously presenting the WHO as an effective institution for risk management. Finally, it was asserted that this task must occur continuously due to the constantly changing nature of the pandemic. However, as will be demonstrated in greater detail below, this narrative was contradicted with regard to the actual discussion of epidemiological information. This is because, as illustrated by co-productionist theory, scientific uncertainty is necessarily embedded in the management of contemporary global risk. Though the WHO attempted to position itself as responsible for minimizing this uncertainty, the absence of solid and incontestable scientific 'facts' surrounding H1N1 was evident. In this way the WHO simultaneously embedded the concept of scientific uncertainty in its risk discourse.

The 'evolving' threat and the uncertain future

The persistent use of the concept of scientific, and particularly epidemiological, uncertainty was prominent in the WHO's account. The mobilization of uncertainty was a significant aspect of the risk discourse. From a co-productionist perspective, the concept of uncertainty can be understood as a boundary object (Shackley and Wynee, 1996), where different groups associate with the concept in different ways. Thus a statistician will utilize 'uncertainty' in one specific manner whereas a policy-maker will conceptualize 'uncertainty' in another. Nevertheless, the idea of 'uncertainty' is ubiquitous in public discourse and (perceived to be) understood by all actors, so that interaction between these actors surrounding and utilizing the concept of uncertainty can persist unchallenged. Such boundary work is necessary in the context of scientific uncertainty because it serves to smooth over potential contestation between competing scientific models. This helps to reconcile the disconnection between scientific uncertainty and scientific claims to authority (Gieryn, 1983; Kleinman & Kinchy, 2003; Lamont & Molnar, 2002). This shared and uncontested, yet group-specific, utilization of a concept is what makes a successful boundary object. In reference to the concept of uncertainty, there is ample evidence to support this argument, suggesting that uncertainty is differently utilized by different actors and groups. For example, the term can be differentially used to denote risk (in the technical sense, quantitative chances of an event); ignorance (where it is not known what information is missing); or indeterminism (where not all of the variables can be understood) (Hansson, 2005; Shackley & Wynne, 1996). Nevertheless, though the concept of 'uncertainty' is applied differently (and often in an ill-defined manner) by different epistemic communities, it is taken to be 'understood' across contexts. Such boundary objects make communications possible at the science–policy interface.

With regard to H1N1, it is indeed theoretically fruitful to conceptualize 'uncertainty' as a boundary object. It is clear that the different actors enrolled into the actor network of H1N1 use 'uncertainty' and also 'risk' (which can be similarly treated as a boundary object) in different ways. This chapter demonstrates that the WHO narrative tended to conflate 'lack of information' with risk and uncertainty in a way that strengthened the organization's argument that it was necessary to adopt extensive measures against H1N1. Later it will be shown that the Council of Europe understood this same lack of information as evidence for the absence of a 'real' risk/threat. The contestation surrounding what had once been a taken-for-granted idea (pandemic risk) is one indication of the way in which H1N1 as a whole became a site of controversy. The WHO's particular characterization of uncertainty was part of its attempt to persuade other actors that H1N1 was a legitimate risk. As a boundary object, the concept of uncertainty can be manipulated to suit various contexts.

Here, the term uncertainty was utilized to meet institutional objectives. Given the multiplicity of scientific knowledges surrounding risk, it is often the case that ideas such as 'uncertainty' become mobilized for particular institutional purposes. For example, in a comparative study of issues of climate change and ozone layer protection in different political contexts, Grundman (2006) finds that scientific consensus is not necessary for policy-making surrounding risk. In some contexts, uncertainty was cited as justification for non-action (e.g. global warming in the USA), whereas in others the precautionary principle was cited as justification to enact regulation where there was a lack of (or conflicting) evidence. In yet other cases, there was a lack of action despite a consensus of scientific opinion. Thus institutionally determined decisions are made in the context of scientific uncertainty (see also Shackley & Wynne, 1996; Srader-Frechette, 1993; Weingart, 1999). In the case of H1N1, the changing (and thereby fundamentally uncertain) nature of the situation was mobilized discursively as a way in which to strengthen the WHO's claim that action needed to be taken against the pandemic.

Through assertions of the constant need for information, the WHO's discourse of risk and uncertainty was linked with another prominent theme that was present in the WHO texts, which was the notion of 'evolution'. Through the use of this narrative technique, it was made clear that a stable and persisting scientific knowledge surrounding H1N1 could not be developed. One of the most clearly repeated terms throughout the WHO documents was that of 'evolving' or 'evolution'. This was employed with regard to almost all aspects of the threat – the nature of the virus, the epidemiological pattern and the general projection of the future progression of the threat. This suggestion that 'The picture continues to evolve...' (Fukuda, 27/04/09b) was commonly repeated throughout the texts. The general impact of the WHO's use of the concept of 'evolution' is illustrated below:

> I think that from the beginning of this whole situation a few weeks ago, we have said over and over, that the situation is evolving and we really don't understand how things are going to go in the future.
>
> (Fukuda, 07/05/09)

Thus, through reference to an evolving situation, it was made clear that there was a lack of control over the future course of the threat. The risk of H1N1 was therefore represented as ever-present, and something to be constantly monitored.

In terms of the nature of the virus itself, it was often noted by the WHO that influenza viruses possess the capacity to mutate, though the term 'evolve' was generally employed instead of the more scientifically common 'mutate'. For example, it was suggested that

> influenza viruses as a group of viruses are just very prone to changing. They mutate easily, they evolve easily and so yes, it is quite possible for this virus to continue to evolve. So when viruses evolve, clearly they can become more dangerous for people, that is to cause more serious disease, or they are also able to mutate so they cause less serious disease and that is very difficult to predict.
>
> (Fukuda, 26/04/09)

The virus was said to 'evolve', constantly changing. This implies that the threat to global health was also inconstant. The risk, in the case of H1N1, was conceptualized as fundamentally unstable; the presumed volatility of the virus served to render it risky.

In addition to the virus itself, its epidemiological impact was also said to evolve. This can be seen in the following examples:

> In terms of the global epidemiological situation, I think it is fair to say that the situation continues to evolve.
>
> (Fukuda, 30/04/09)

> So our overall assessment is that the situation continues to evolve as we have been stressing from the beginning, and in keeping again in the messages from many speakers, we are not quite certain how this is going to evolve.
>
> (Ben Embarek, 04/05/09)

This evolving situation justified the WHO's assertion that H1N1 represented a persisting threat, despite media (and other) questions regarding the validity of the organization's continuing risk discourse:

> Now, if we take a look at where we are right now in terms of the overall global picture, I think there are a couple of things that are very clear. One of them is that the situation in the global picture still continues to evolve. I know that there is much speculation in the media for example, that may be things are over or in some countries it looks like things are going down. But really, from a global perspective and from what we are seeing, this is probably fairly early in the spread of this infection. It is clear that the global picture continues to see spread of this virus and an evolving picture of the epidemiology.
>
> (Fukuda, 22/05/09)

With regard to viral characteristics, disease epidemiology and severity then, the WHO explicitly suggested that the H1N1 threat was in a constant state of flux. This served to reinforce the risk-laden discourse surrounding the pandemic. It also reasserted the positioning of the WHO as an institution that effectively monitored the situation and provided global populations/governments with information regarding such risk.

This was again notable in other suggested 'evolving' aspects of the threat. In response to low morbidity and mortality rates, the WHO emphasized the uncertainty of the situation:

> I think again you cannot make those projections until you really see much more of what proportion of people get seriously sick, what

proportion of people who are infected die. We said over and over again, we are in this period where the spread of the virus is evolving. Our understanding of what the clinical spectrum is, is evolving, more information is being collected. But I think it is very premature to make those kinds of projections.

(Fukuda, 07/05/09)

This allowed the WHO to maintain a risk discourse despite the low impact of H1N1. This was similarly evident with regard to severity:

This picture is changing, and so this is why we have stressed about [sic] the evolving nature of the situation, this is why we have really refrained from jumping quickly to say: 'this is mild', 'this is something', because we know that we are seeing things change on an almost daily basis.

(Fukuda, 11/05/09)

At the beginning of this kind of phenomenon where we are now... there is a great deal of attention being paid to try to figure out what is the severity of this illness, and I hope you appreciate that we have been very careful to say that we expect the situation to evolve, and we are very careful to say that we do not know quite how it will evolve.

(Fukuda, 05/05/09)

This narrative of unpredictability meant that the WHO could counter claims of mildness. Furthermore, even if H1N1 eventuated as mild, the emphasis upon uncertainty that was evident early in events justified the risk discourse. In terms of the potential severity of H1N1 and against criticisms and suggestions that the virus would eventually turn out to be mild, it was asserted that

I would be very pleased if it turns out that this virus is weaker than it could be, I would be the most happy man in the world. However, I think that history has told us that these viruses are very, very, very unpredictable. And this virus is spreading in human populations, these viruses mutate. These viruses change, these viruses can further reassort with other genetic material... However, any evidence that pushes us towards being able to issue statements on less severity will be very reassuring for the world.

(Ryan, 02/05/09)

And

> We want everyone to understand that what we see now is important, but to remember that this is a virus, this is a situation in which things that evolve, and which things can evolve quite differently and that is why quite much attention is being taken to what is going on. This is why we are jumping so hard on it because if it stays mild and people stay healthy, then that is great, that is the best possible outcome. But if it does turn severe, then this is something that we have to know about it, it is something that we have to be prepared for, and it is something that we have to jump on.
> (Fukuda, 05/05/09)

Co-opting the concept of evolution into discourses of risk allowed for the continued construction of H1N1 as (potentially) threatening, even when early indicators suggested a mild event. Moreover, if the virus and its epidemiology were represented to be changing rapidly, then constant vigilance was necessary and the role of the WHO was reinforced.

The concept of evolution was also proposed in the terms of more generalized references to the future. Here, early in events, the uncertain future of the virus was noted:

> I think that a fair question to ask is where we are going. Is it theoretically possible that this epidemic could certainly stop for unknown reasons, although this is probably unlikely at this point. It is also possible that we could continue on with spread of relatively mild illness in most countries recognizing that death and serious illnesses will occur sometimes. And it is also possible, that as we go into the future, we will see more serious cases. These options are all possible.
> (Fukuda, 29/04/09)

This suggestion of constant evolution conjures up images of unpredictability. The WHO's depiction of the future of the H1N1 threat thus heavily subscribed to these notions of risk and uncertainty. The future of the threat was depicted as indistinct and susceptible to unforeseen variations, and this characterized the WHO's discourse about the risk posed.

Statements regarding uncertainty and evolution reinforced the assertion that there was a constant need for new information with regard to the threat. They highlighted the essential institutional role of the WHO as the custodian of this information, and served to protect it if the threat did not manifest, given that the organization's reactions were depicted as being based upon the potential for future changing circumstances. The WHO was placed in the position of needing to construct itself as providing information, while it was simultaneously impossible to gain full scientific consensus or coverage under these conditions of novel risk. Scientific uncertainty was embedded into the nature of the H1N1 virus. This uncertainty was also pivotal to the WHO's role as a risk-managing institution. In this case, the organization co-opted the idea of scientific uncertainty into its risk discourse surrounding the pandemic.

Statistics

In addition to the pandemic state being characterized as uncertain, one intriguing aspect of the WHO's risk discourse relates to its depiction of epidemiological statistics. As suggested above, one of the assertions made by the WHO in relation to risk was the suggestion that the continual collection of information was necessary to combat uncertainty. However, and somewhat paradoxically, the WHO simultaneously depicted epidemiological statistics as themselves fundamentally uncertain. This illustrates the deep integration of scientific uncertainty into the science–policy interface that moulds the problem of H1N1.

Commonly the use of statistics is one pervasive method through which medical risks are brought into being. In Western societies statistics are often treated as objective representations of the truth (Best, 2001). In contrast with this perception, due to their nature as 'facts' which are primarily constructed by people, statistics are necessarily subjective. For example, as has been argued, 'risk' is not an objective concept but is rather in itself a socially produced and variable notion. In this way, risk statistics are wholly dependent upon statisticians' conceptualizations of what 'risk' actually constitutes (Bartholemew, 1995; Gigerenzer, 2002). Furthermore, according to Hindess (1973) and others, even more critical is the fact that statistics are produced through collective activity and are therefore tied to the specific organizational context and the overall cultural meaning system in which the construction is embedded (Hacking, 1999; Hindess, 1973). The WHO's use of statistics in the case of H1N1, however, does not follow the common model of scientific fact-making through reference to a statistical/empirical 'reality' (in the positivist sense). Rather, as with the suggestions of 'uncertainty' and 'lack of information' above, statistics were themselves presented by the WHO as uncertain. Again, the very lack of a consensus upon statistics was represented as a matter of concern, and this perpetuated notions of an unstable future.

Statistics were presented by the WHO as having limited value in evaluating H1N1. In fact, the organization's representatives pointed to the inadequacies in comparative statistics in order to quash the notion that H1N1 was a mild disease. For example, in response to a journalist's suggestion that the death toll for H1N1 was lower than that of seasonal influenza, Fukuda responded:

> So the reason why this has been confusing is that with the current pandemic situation, the numbers of people that have been reported to die from [pandemic] influenza are people in whom our direct testing has been done so someone has taken a sample and sent it to a laboratory. This is not usually how we count deaths from influenza.
>
> And when we do that [use the same statistical method in calculation as for seasonal influenza] I think that we will find that in fact the number of deaths worldwide is much higher than the 10,000 [as was currently reported for H1N1].
>
> (Fukuda, 17/12/09)

Statistics were therefore depicted as inadequate and of limited value in assessing risk. For example, it was stated that the use of statistics to compare seasonal and pandemic influenza was inaccurate.

The question of the statistical methodology for calculating the impact of disease also surfaced in reference to comparisons made between H1N1 and previous pandemics. Several of the media questions centred on allegations of the mildness of H1N1 in comparison with historically experienced pandemics. They suggested that H1N1 was not, comparably, a 'real' pandemic threat. In one instance, Helen Branswell (Canadian Press) stated that

> there are certainly people who feel that this has turned out to be much less of a threat to global health than was first thought. In fact CDC [US Center for Disease Control and Prevention] has come up with the fatality ratio of 0.018, which people are pointing to suggest this is quite a mild event, and I am wondering if you could address... that notion that it's very mild...
>
> (Fukuda, 03/12/09)

In addition, the reporter Richard Knox cited the same CDC calculation as Branswell, stating that the 'CDC has been calculating a case fatality of 0.018 percent which is about 100 times less severe than the 1918 case fatality rate' (Fukuda, 03/12/09). Here, as above, the answers given tended to make reference to the technical details of such calculations to indicate the invalidity of these comparisons. Thus

> WHO has not calculated case fatality rates based on current information. But there is a reason for that and it is an important reason to understand.
>
> With the current pandemic, we really have data which is almost an anomaly, when we look at how influenza has been counted in the past... [by illustrating data collection methods] and so I think that it will take another one to two years, after the pandemic, for those data to be collected and for the kinds of estimates which are typically done for seasonal influenza, but also for pandemic influenza, to be done to make the estimates.
>
> I think, that when we have those estimates, then we will be in a much better position to really talk about how does this pandemic [sic] stack up with the earlier pandemics, as well as with seasonal influenza.
>
> (Fukuda, 03/12/09)

In this way, suggestions of the mildness of H1N1 were deflected by assertions that the statistics were irrelevant to the situation at hand. This may have arisen as a response aimed at minimizing conceptions of mildness and emphasizing risk. However, it was also evident that the representatives portrayed statistics as generally problematic, and to some extent demonstrated the scientific uncertainty surrounding the problem of H1N1.

It is apparent throughout the texts that the WHO's representatives were determined to depict any focus upon numbers as unjustified and unhelpful. The uncertainty of statistics was thus a continually reiterated theme. This is seen in the examples below:

> One of the first things I want to caution everybody about is that we are in an evolving situation so we cannot be too focused on numbers. The numbers can change quite rapidly as we know from any outbreak situation.
>
> (Fukuda, 26/04/09)
>
> As you know, we have been really stressing the fact that we shouldn't focus too much on the figures because they are pretty fluid and they can change fairly often.
>
> (Ben Embarek, 04/05/09)

Not only were the statistics depicted as fluid and changing but they were also suggested as being irrelevant:

> as we go into this situation, the numbers themselves will become a little bit more irrelevant. We now have countries that are moving away from counting cases individually because there are too many cases. So just to give you [a] heads up, we will begin to de-emphasize the numbers because they will increasingly not reflect what is going on.
> (Fukuda, 22/05/09)

As these quotes show, the WHO's official line of argument suggested that morbidity and mortality rates should not serve as a basis for assessing the state of the pandemic. Statistics were depicted as uncertain, and only represented the uncertain, evolving situation.

It is clear, however, that media questions at press conferences tended to dwell upon statistics. Specifically, many reporters often queried discrepancies between different institutional sources of statistics. Given the uncertain depiction of statistics, the WHO's representatives reacted to questions about morbidity and mortality rates by attempting to deflect them. Statistics were repeatedly depicted as problematic, and difficulties in collection were cited as the reason why statistics should not be focused upon and why discrepancies existed. Thus

> In terms of the different numbers of deaths, I think that one of the features that is simply to follow investigations especially when you have big outbreaks occurring, is that the numbers can be very confusing and you can have cases of disease reported, cases of deaths reported, and then some of them might be laboratory-confirmed deaths, and other times these deaths which are epidemiologically suspicious but not laboratory-confirmed. I cannot address directly why do the numbers vary a little bit, right now, but I do know how these outbreaks unfold and how difficult and overwhelming it is to get the numbers quite straight. It is very common to have the numbers vary somewhat in the beginnings of these large outbreaks. At this point, I cannot address the specifics, but that is generally, what is true with the outbreaks.
> (Fukuda, 29/04/09)

Statistics were problematic because they were inaccurate. Again in terms of mortality rates,

> This is a figure that we do not track very carefully. The suspected cases – all national authorities investigate disease cases – and then

there are ones that they have confirmed cases by doing the laboratory testing and then they have other cases that they are looking at. But is it not something that we ask the countries to report to us, it is not something that we track because laboratory-confirmed cases is really the clearest way to monitor the spread of the virus around the world. Then we are not dealing with ambiguities, I simply don't have those figures, I can't tell you how many investigated cases there are now.

(Ben Embarek, 04/05/09)

There was great uncertainty in this regard, and the best global estimates were given in terms of the extrapolation of the statistics of individual member states:

In terms of flu deaths, in purely epidemiological estimates of the number of any deceased, you certainly know that there is a big uncertainty. The surveillance is not that precise, so therefore, you need to take a number of countries and then you do modelling, and then you extrapolate.

(Fukuda, 06/05/09)

It is suggested therefore, that the statistics were uncertain because their collection was the responsibility of national governments, and the variability in collection methods rendered such statistics incomparable.

Thus, in the WHO's account, statistics were depicted as an unreliable entity. Nevertheless, at specific points in the texts, the WHO's spokespeople acknowledged that some numbers are important. This occurred for example in relation to phase changes (see Chapter 5). In this instance it was stated:

So out of the thousands or so cases which people have been bandying about in public fora, really the number of cases we are sure of is much much smaller. So before we change phases it would have been irresponsible of us not to understand the nature of the outbreak better before we changed phases.

(Härtl, 27/04/09)

Furthermore, in some instances the spokespeople sometimes cited statistics. For example, at the start of the press questions in one conference, Fukuda replied to a question about potential mortality rates:

If you take approximately two billion people – that would be the third of six billion people – it just means that, if you have a virus that is capable of leading to serious illnesses, again you can have very large numbers of people getting sick and requiring hospitalization.

(Fukuda, 07/05/09)

But then at the end of the conferences, to clarify the point, he added:

I want you not to walk out of here saying that there is an estimate of 2 billion people to get infected over the next year or so. What I am pointing out is that in the past, when we have had pandemics, approximately about a third of people have gotten infected, but again in keeping with all things about the future, we live in a different world. This is a benchmark from the past, so please do not interpret this as a prediction for the future.

(Fukuda, 07/05/09)

Such an example suggests the confusion regarding the use and importance of mortality rates. It also reinforces the problematic nature of attempts to allude to scientific 'facts' in the WHO's policy and discourse surrounding H1N1. The organization characterized the science surrounding H1N1 as itself uncertain, and in this way opened up its decision-making to scrutiny as not having been made on the basis of 'objective facts'.

The WHO therefore failed to definitively establish the role of statistics in assessing H1N1. Throughout the pandemic it was asserted that epidemiological statistics represented an important source of information in combating risk and uncertainty. Such an outlook represents a standard epidemiological approach to the understanding of statistics – that mortality and morbidity statistics represent the objective scientific reality of disease. However, simultaneously, the statistics themselves were represented by the WHO as uncertain and meaningless. In effect, the organization attempted to emphasize the variability in the construction of statistics in response to the context of uncertainty in which it was acting. Nevertheless, given that epidemiological statistics are commonly assumed to be indicative of the reality of disease severity, the WHO's depiction of statistics lacked stability both in terms of its contradictory nature and in its failure to subscribe to dominant scientific notions of statistical objectivity.

The paradoxical depiction of the importance of statistics relates more generally to the evidence that the WHO failed to produce a convincing risk narrative in the case of H1N1. To demonstrate the need for sustained interest and effort against the pandemic it was necessary for the organization to strongly characterize it as a risk. However, the texts under analysis presented inconsistent depictions of risk through an incoherent usage of the associated term of severity. It was shown that the concept of severity underwent a number of changes as a reaction to the (mild) progression of the pandemic. This included attempts to render the concept increasingly complex and even to abandon the concept altogether. As with the definition of pandemic illustrated in Chapter 2, the WHO's depiction of 'severity' demonstrates a failure to effectively mobilize a consistent discourse of risk. In this way the overall construction of H1N1 was rendered vulnerable to contestation.

Theoretically speaking, institutionally determined decisions regarding risks are made in the context of scientific uncertainty and competing scientific discourse. Nevertheless, the WHO, as the risk-managing institution, needed to evaluate and accept one consistent model of risk/threat in order to present a stable translation of H1N1 as a pandemic risk. However, after adopting the geographic explanations of severity as equivalent to risk, the subsequent unfolding of the situation uncovered the tenuousness of the initial assumption. This eventually resulted in the contestation of the concept of H1N1 as a whole. Thus it is clear that 'Those responsible for scientific policy occasionally run the risk that a piece of unanticipated reality may be lurking behind the metaphorical imagery they have constructed in order to accommodate a broad spectrum of ideas' (Shrader-Frechette, 1993: 63). Institutions must choose from a range of scientific possibilities, and if the predicted model does not manifest in reality then the decision-makers are open to criticism. The WHO attempted to use this as a defence, emphasizing that it was making decisions in the context of scientific and statistical uncertainty. Some commentators argue that the notion of uncertainty 'can serve as an alibi in accounting for a lack of policy effectiveness...' (Shackley & Wynne, 1996: 277). However, references to such uncertainty did not ultimately help the WHO in the case of H1N1. This is because, after the uncertainty had passed, the once uncertain future seemed obvious in hindsight. Given that science is always presented as complete and unanimous, the decisions of the managing institution retrospectively appear to have been mistaken, and as such the WHO was ultimately held liable for declaring a 'false pandemic'.

4
Categorizing H1N1 – The Pandemic Alert Phases

A key aspect of the WHO's pandemic management strategy is the organization's official definitions of pandemic categories. This chapter illustrates the WHO's attempt to define the concept of 'pandemic' through its Pandemic Alert Phases. H1N1 was technically categorized as a 'pandemic' as a result of this official classificatory schema. The act of categorization was central to the controversy surrounding H1N1. The widely perceived miscategorization of the disease rendered the concept of 'pandemic', and the WHO itself, open to critique. Sociologically speaking, this can be explained by the powerful role of categorization within social life, and specifically within scientific debate. In theorizing the importance of classifications in the case of H1N1, this chapter provides an account of the function of categories, demonstrating that the WHO's failure to effectively produce a robust definition of 'pandemic' within its phases was essential to the contestation over H1N1. The ill-defined phases were not singularly a result of the WHO's institutional processes. Rather, the very idea of phase definitions was predisposed to vulnerability, given that the concept 'pandemic' was an indistinct boundary object. Furthermore, the scientific and institutional structures surrounding the phase categorization tended to produce simplistic definitions, which cannot adequately reflect the complex manifestations of disease.

To appreciate the significance of the phase definitions, it is necessary to first give a brief overview of their use within the WHO. The Pandemic Alert Phases are the official set of WHO definitions surrounding the level of preparation that is necessary to combat a potential influenza pandemic threat. They serve as a signal of the pandemic potential of any circulating viral strain. Member states then react to this signal in formulating management strategies. The phases are set to reflect the estimated probability of a pandemic, with Phase 6 indicating a pandemic in progress (WHO, 2009: 27). The organization's overarching narrative about its definitions was that 'The phases are applicable to the entire world and provide a global framework to aid countries in pandemic preparedness and response planning' (WHO, 2009: 24). Phase declarations therefore act as an important indicator of risk and a method through which to distribute key information and conceptualizations of pandemic threats from the WHO to its member states.

The pandemic phases were redefined just prior to the first recorded incidence of H1N1. This redefinition was outlined in the updated version of the organization's 'core document' regarding influenza management, the *Pandemic Influenza Preparedness and Response* (2009) guidance document, which was produced by the Global Influenza Programme. Here the organization outlined the phases, asserting that the 2009 definition 'Retains the six-phase structure [from the earlier 2005 version] but regroups and re-defines the phases to more accurately reflect pandemic risk and the epidemiological situation based upon observable phenomenon' (WHO, 2009: 3). However, the WHO's new definitions were poorly developed and contributed to its difficulties in the management of H1N1. As will be demonstrated in Chapter 6, this redefinition of the phases immediately prior to the incidence of H1N1 presented one avenue for the condemnation of the WHO by other global actors.

In respect of H1N1, member states were confused by, and reacted with criticism to, the WHO's conceptualization of the pandemic phases. From the early stages of the H1N1 threat there was an intense level of scrutiny surrounding the phases, both in terms of the WHO's categories and in the context of its timing of phase increases. In addition to the consternation surrounding the 2009 redefinitions, the practical implications of the phases produced confusion; the WHO and its wider audience (here, particularly member states) adopted divergent interpretations of the implications of the phase categories. This contestation will be covered in depth in Chapter 6. Here, the WHO's perspective and discourse surrounding the phases will be focused upon. It is argued that, combined with the WHO's failure to produce a consistent and robust risk discourse, the failure of the WHO to have effectively created a common understanding of the phases resulted in a wider definitional problem for the organization. This led to widespread criticism and eventual re-evaluation of the phases. The controversy also undermined the concept of an H1N1 pandemic as a whole and weakened the WHO's claims to legitimacy.

The controversy surrounding the phase definitions can be explained through a reflection on the sociological importance and impact of classificatory schemes. Classification is intrinsic to thought and social structures (Foucault, 1970; Lewin, 1994; Moscovici, 1988). As such, a great deal of classical and contemporary sociological work deals with the problem of classification, either explicitly (in attempts to understand classificatory schema) or implicitly (in themselves producing methodological or theoretical classifications of social phenomena). The more explicit sociological theorization regarding the role of classifications will be discussed here in order to explain the impact of the WHO's classificatory scheme upon the construction of the H1N1 pandemic. Classifications are socially significant in a number of ways. In theorizing or conceptualizing a novel phenomena, there is a tendency to think in terms of analogies to phenomena which are already socially understood, recognized or defined (Douglas, 1969, 1973; Friese, 2010; Sontag, 1978). Institutionalized classificatory schema, such as the WHO's Pandemic Alert Phases, are a way in which this type of analogizing is formalized (Lewin, 1994). Placing phenomena or ideas into a classificatory category is thereby a key epistemological strategy, which allows for understanding a phenomenon in relation to another which is already 'known' (Lewin, 1994; Martin, 2004; Martin & Lynch, 2009).

The success of a classificatory scheme lies not in its correspondence to some external objective realty but instead in its correspondence with the discourse of the thought collective which created it. This concept is well expounded in Foucault's (1970) example of the classification of imaginary beings, or Douglas' (1969) example of the classification of animals according to the book of Leviticus. These two examples show the function of categories in upholding existing social boundaries. In the context of Western scientific theory, Kuhn and Fleck demonstrate that classificatory knowledge is a product of the prevailing paradigm, discourse or disciplinary thought-style (Fleck, 1979; Kuhn, 1970). From this it is clear that the classificatory schemes which are thought to explain the world are actually a method of constituting it (Freeman & Frisina, 2010; Lewin, 1994; Vaihinger, 1949). Nevertheless, on occasions where classifications are constructed to serve a practical (rather than a purely theoretical) purpose, the schema must also fulfil its functional role. In the case of the pandemic phases, this role was to assist member states in recognizing and managing a pandemic threat. Thus although, as the constructionists suggest, classifications can always be differently constructed, in the case of the phases it was critical that the WHO chose a construction which was both robust and fulfilled its functions.

The WHO's pandemic phases became a site of considerable controversy because they did not accomplish the circumscribed functional objectives. I argue here that while categories are socially constructed, they can simultaneously serve to fill a functional purpose. Those theorists who emphasize the functionality of classificatory schema (e.g. Durkheim and Mauss (1963) classically, or Douglas (1969, 1973)) provide a convincing argument for the functional role of categorization. Such theorists emphasize the importance of classifications in circumscribing social roles and maintaining social boundaries and order. However, a theoretical emphasis upon functionalism can only partially explain the utility of classifications. Constructionist approaches also provide insight, suggesting that classifications in themselves produce the social 'facts' which they claim to elucidate. Such theorists emphasize the often arbitrary and power-laden nature of classifications in the production of social reality (Haraway, 1991; MacKenzie & Wajcman, 1999; Treacher & Wright, 1982). In explaining the case of H1N1, a medium between these two positions can be reached – the category of 'pandemic', while a product of social and institutional forces, attempts to explain an observable phenomenon for a specified purpose.

In the context of scientific knowledge production, such as defining the concept of 'pandemic' and categorizing its 'phases', the act of classification is pivotal to understanding 'natural' phenomena. Scientific knowledge fundamentally consists of classificatory schema. These are recognized within scientific disciplines to mirror objectively defined differences or similarities in the natural world. Debates about classification are therefore at the centre of scientific work. In fact, shifts in scientific thought can be conceptualized as the outcome of the continuous revision of classificatory schema. However, this is not meant to imply that there is some form of 'best' or 'correct' form of classification – classifications often reflect a particular institutional or disciplinary perspective (Fleck, 1979; Foucault, 1970; Lewin, 1994; Martin & Lynch, 2009). Nonetheless, schemes can be either functional or non-functional with respect to the purposes for which they were created. In the present example, the phase categorizations are not conceptualized here as somehow objectively 'incorrect'; rather, they did not serve the function for which they were produced, which was to aid member states in assessing and managing their risk, and thereby reduce uncertainty. Through this, an implicit role of the phases was also to legitimize the role of the WHO as arbiter of pandemic events, which (as this chapter demonstrates) was another function which was not effectively fulfilled.

When considering the use of classification in the natural sciences, such schemata do indeed exhibit a type of functionality in terms of discipline-bound knowledge production (Bowker & Star, 1999; Dupre, 2006; Swistak, 1990). As such, the pertinent question concerns the relationship between the phenomenon observed and the classificatory scheme which assumes to convey knowledge surrounding it. In the case of H1N1, its characteristics were fitted into the classification of a Phase 6 'pandemic' but diverged too far from the implicitly understood (black-boxed) conception of what a pandemic constitutes. As a result, both the classificatory scheme and the WHO's credibility in relation to influenza pandemics came under contestation. This is because, though a social/political construction, the WHO's phase definitions failed to fulfil its function.

The discipline-bound production of categories serves to simplify the explanation of phenomena and eliminate other potential competing accounts (Freeman & Frisina, 2010; Lewin, 1994). This is done through judgements of what constitutes normal/standard characteristics within a given category and where the boundaries between different categories lie (Derksen, 2000). Within the sciences, every classification is a translation from a complex natural phenomenon (the spread of the H1N1 virus) into the conceptual scheme which is available to represent it (the Pandemic Alert Phases). It is also a collective process which is socially validated by the scientists who are working on the problem (here, the WHO expert panels) (Freeman & Frisina, 2010). To make judgements regarding both the classificatory scheme and the phenomena which fit into it, the descriptions of the natural world need to be simplified and unified. This process also eliminates a plurality of potential perspectives born from other thought collectives or disciplinary communities (each of which would be adequate in its own terms) (Jasanoff, 2004a; Lewin, 1994). Through such schema, phenomena either get bound together (where differences are ignored and similarities emphasized) or rendered distinct (where differences are exaggerated).

To be able to place any specific phenomenon within a classificatory scheme, it is also necessary to clearly define what counts as a member of a particular category. Such placement is thereby 'an *achievement* in a field of alternate epistemic categories' (Martin & Lynch, 2009: 246). Placing H1N1 into the category 'Pandemic Phase 6' involved a set of judgements surrounding what that phase meant. It also involved an act of rendering the object of the H1N1 virus docile (Martin & Lynch, 2009) and simplified. As has been demonstrated, the virus had to be (re)constructed as a pandemic strain in order to be counted as such. This demonstrates the effect of the WHO upon the constitution of H1N1. The virus could only have become a 'pandemic' as a result of the institutional act of categorization. However, the WHO failed to properly translate H1N1 into the category 'pandemic', leaving both itself and the concept of 'pandemic' open to criticism. This was most evident at both the start and the end of the pandemic event, as discussed in detail in this chapter.

The presumed function of the Pandemic Alert Phases

The WHO's pandemic phases set out to distinguish the degree of preparation needed in reaction to any influenza strain that threatens to develop into a pandemic. The phases classify what does and does not constitutes a concern by outlining the factors which define a pandemic, or (in the earlier phases) a potential pandemic. Once the categorization (the alert phases) had been formed, any specific phenomenon in question (here, the H1N1 virus) is measured against the predefined criteria. That is, the H1N1 virus was epistemologically constituted as a pandemic at the point where it fitted into the WHO's category of Phase 6 pandemic. The classificatory scheme is therefore not purely theoretical, as in some scientific work, but also performative (Austin, 1980; Bowker & Star, 1991; Friese, 2010; Martin & Lynch, 2009). At every phase in which the virus was classified, national governments were expected to enact preparatory measures which correspond to the level of threat. In this way the WHO's phases provided a performative discourse which constituted an important interface between meaning-making and action.

The phases evoked certain types of action because the categories serve to define the virus through classification. Once a classificatory scheme has been produced, it tends to acquire a taken-for-granted nature, where the categories are presumed to be 'natural'/'true' (Douglas, 1969; Foucault, 1970; Martin & Lynch, 2009). Although intended primarily as a description, the phases also implicitly predict and explain the nature of pandemics. This is because it is impossible to describe merely what constitutes a pandemic without simultaneously imparting conceptualizations of the nature of 'pandemics' as a category state. This act of definition therefore foreshadows what constitutes effective action against a pandemic. As such, to ascribe a virus to the category 'pandemic' is also to constitute it. Phenomena are only constituted as ontologically distinct when they are rendered classifiable (Latour, 2005; Martin & Lynch, 2009). Thus H1N1 became a global matter of interest and concern as a consequence of its classification as a pandemic.

In turn, this classification was itself a result of the construction of the initial classificatory schema – the alert phases. As well as placing H1N1 into the category of 'pandemic', it also distanced the virus from previous (or other potential) conceptualizations.

In the perspective of many other actors, H1N1 did not represent a genuine pandemic threat. However, when it was placed by the WHO into the category Phase 6 pandemic, it was therefore constituted as such in terms of both discourse and policy. The reason for this was two-fold. First, as has been demonstrated in previous chapters, the translation and conceptualization of the H1N1 virus and its associated risks was fragile. Second, the WHO's Pandemic Alert Phases were themselves not definitionally robust. The WHO claimed that the phases described states of risk and indicated recommended reactions to a pandemic threat. However, on the whole, the characteristics which were chosen to measure the phases did not fulfil this aim. To demonstrate this it is first necessary to illustrate the way in which the organization conceptualized the categories.

One important aspect of the WHO's depiction was the way in which the phases were created. That is, it was strongly maintained that the pandemic phases were constructed through a collaborative process of scientific experts. Here it was suggested that

> The phases are the result of a consensus between WHO and its Member States. It is the result of a number of technical consultations where Member States were invited to discuss and to conclude that the phases that are currently in the *Pandemic Preparedness Guideline* are probably the best way to approach this phenomenon.
> (Briand, 08/05/09)

And again, in relation to the revised 2009 version, it was suggested by the WHO that

> In terms of the WHO Pandemic Alert Phases... WHO has worked with countries and scientists over the past two years to really improve [them]. They have been around for a while... going back to the late 1990s, but in an effort to update the guidance to Member States, what we did is to sit down with a very large number of scientists and public health people from around the world...
> (Fukuda, 11/05/09)

As these statements demonstrate, the WHO texts suggested that the organization was not solely responsible for producing the phases. By citing collaboration with member states and scientific experts in the process, the WHO diffused the responsibility for the outcome. Furthermore, as suggested by co-productionist theory, this very plurality of perspectives and users may have resulted in the (over)simplification of the phases, to the ultimate detriment of their utility (Jasanoff, 2004b; Shackley & Wynne, 1996; Shrader-Frechette, 1993). In the presence of competing perspectives, the WHO had to favour one above others, and simplify real-world events in producing an all-encompassing classificatory scheme.

The utility of the phases reflects the WHO's position within global public health. Its role is to monitor disease threats and to determine phase increases as deemed necessary. National governments then act as a reaction to the WHO's announcements. In this way, accounting for phase increases is pivotal to the task of the organization in pandemic management. However, for the WHO, these declarations did not in themselves determine the actions of member states. The WHO asserted that it was not predominately responsible for the actions that were taken by member states in reaction to its announcement of phase increases for H1N1. For example, it asserted that 'these are tools – the Phases are tools really to help countries in their efforts to be ready for pandemic influenza...' (Fukuda, 26/05/09). The phases are suggestive of preparatory actions, but from the WHO's perspective the responsibility lies with the national governments as to what a specific phase change might indicate for their particular national conditions. Thus both descriptions of the production of the phases and descriptions of the meaning of the phases served to minimize the responsibility of the WHO and to obscure the function of the classification. This obfuscation in part weakened the functionality of the phases.

Furthermore, while the WHO distanced itself from the responsibility for the phases, other aspects of the categorization were also unclear. One important area is that of risk and severity. In some instances the WHO acknowledged that the phase categories do indeed serve as a measure of pandemic threat. In the context of phase increases,

> it was also felt that given the rapidly developing situation, that it was important to send a strong signal to countries [by increasing the phase], that now is a good time to strengthen preparations for possible pandemic influenza...
>
> (Fukuda, 27/04/09b)

In this quote it is evident that the phase changes are conceptualized as a signal made by the WHO to member states to perform preparatory actions. Phases are also conceptualized here as corresponding to levels of threat.

Nevertheless, the WHO narrative surrounding phases simultaneously (and counterintuitively) also served to suggest that they did not represent a clear indication of the nature of the pandemic severity. Here it was proposed that

> One of the things to understand about pandemic Phases that is really important, is that these Phases are not intended to be a barometer of epidemiology per se. This is not a measurement of epidemiology per se, but it is really a warning and an alert to countries and to the global populations, that the risk of this new virus spreading and reaching their countries is now judged to be significantly higher and it is really a call for governments and people to really take stronger preparations, to move ahead and take the preparations that they need in order to reduce the health impact of the new virus.
>
> (Fukuda, 30/04/09)

Here the transition through phases is asserted not to express an epidemiological reality but to 'alert' governments regarding spread. The epistemological shift that is occurring here is a transformation from the 'pandemic' state that signifies a risk of severe disease to signifying an appraisal of geographical spread. Using these examples it is clear that movement through the alert phases was suggested to serve as a signal from the WHO that member states should escalate their preparations, but it is equally unclear as to how (or if) the phases measure risk in the sense of epidemiological impact. Thus the organization failed to present an adequate representation of practical actions to be taken, undermining the presumed utility of the phases.

The quote below further illustrates these contradictions in meaning, responsibility and function, and particularly the problematic concept of 'severity':

> I would just like to speak a few words about what a possible move to Phase 6 [full pandemic phase] might mean. Pandemics are serious, but it is important to know that Phase 6 describes geographical spreads of the disease not its severity. We do not know how severe or mild this pandemic might be. History has shown us that disease activity in the past pandemics is like a patchwork, while every country is ultimately affected, the state of the development of the epidemic in any given country can be very different at the same time. Therefore

measures taken by governments will differ, depending on the state of the development of the epidemic in that country. This is the time for us to prepare and be ready.

(Ryan, 02/05/09)

This extract suggests that it is geographical spread which represents the quality of the threat that is being measured, as opposed to severity (risk and impact). This means that the proposed link between phase increases and increased preparation is highly tenuous. Thus, in this context (and as has been presented in Chapter 3) the inability to demonstrate severity/risk in a robust manner made the WHO's account highly problematic. Furthermore, the utility of the phases was minimized by the suggestion that they might not reflect local conditions. Taken together, this formed the basis of questioning and criticizing the phases.

Thus the WHO's definition of the phases revolved around descriptions of the geographical spread of the virus and the implied susceptibility of populations, overlooking severity. Given that the pandemic phase definitions were produced by the WHO before the H1N1 threat, this discounting of severity within the classifications was not a result of the organization attempting to 'create' the scare. A more persuasive explanation can be found with reference to the organizational and social structures in which the phases were produced. One explanation for overlooking severity may be found in the climate of risk perception surrounding infectious disease threats in which the current phases have been established (Abeysinghe & White, 2011; Brown et al., 2009; Eichelberger, 2007). This is evident in the following statement:

> Just to go back to the whole reason for why we have the Phases, why we have been working on pandemic preparedness, as you know one of the main dangers that we have been focussed on through the past three, four, five years, has been H5N1 and avian influenza. This is really something that drove the pandemic preparedness process very hard, and that was one of the reasons why we have worked pretty hard to update guidance and clarify the Phases. And again, it is not just H5N1, but I think one of the lessons from SARS and the other global outbreaks that we have dealt with over the last number of years...
>
> (Fukuda, 11/05/09)

While both H5N1 (avian influenza) and SARS were never declared pandemics, they were respiratory illnesses of a similar nature and similarly perceived as threatening. The current response to H1N1 was framed in the context of the experience of these two past threats; it was clearly suggested that the 2009 phases were modelled in part as a response to these. Notable here is that both H5N1 and SARS were not declared pandemics because those threats' geographical concentration (for SARS) and their inefficient mode of transmission (for H5N1) excluded them from the classifications. However, they were similar to each other in that they both resulted in high rates of morbidity (i.e. when an individual did contract one of these viruses, their chances of severe disease or death were high). Both H5N1 and SARS were severe diseases in the black-boxed sense, whereas H1N1 was characterized by low morbidity and mortality, and thereby 'mild'. It can be argued that to some extent the phases left out the notion of severity because it was not a contentious characteristic in the WHO's previous experience of influenza threats – severity was assumed. Nonetheless, the exclusion of severity as a defining characteristic of the phases was a key oversight, and it decreased the functionality of the categorizations as a whole.

During the process of defining the pandemic phases, the severity of a pandemic was taken for granted and other definitional factors were focused upon. Overlooking severity was significant because this resulted in a classificatory problem for the WHO in the case of H1N1. Here the technical characteristics of a Phase 6 pandemic (as measured by spread and novelty) were present, but impact – that is, severity – was not. This eventually resulted in both the construction of the H1N1 threat and the phases themselves being made vulnerable to criticism.

Heightening alert phases and declaring a pandemic

The criticism of the pandemic phases revolved around two (seemingly contradictory) questions which arose at different temporal points in the event. During the later stages of H1N1's spread, questions revolved around whether a pandemic declaration had been necessary and, finally, when a post-pandemic period would be announced. However, during the early predeclaration period, critics in fact argued that phase increases were occurring too slowly. Both of these criticisms can be explained through the imprecise nature of the phase categories, which rendered them liable to diverse interpretations and diminished their functionality.

A key function of scientific classifications, particularly in the context of risk, is to erase uncertainty (Derksen, 2000; Martin & Lynch, 2009). Translating a phenomenon into a classificatory category constitutes it in accordance with the category's defining characteristics. So classifying H1N1 as a Phase 6 pandemic constituted it as a pandemic disease. Thus, when effectively employed, classifications eliminate ambiguity by definitively placing the phenomenon into a class of things. (In this case the uncertainty to be abrogated was the question surrounding how the virus should be managed). This act of erasing uncertainty is central to scientific practice. To be constituted as a scientific 'fact', an idea/phenomenon needs to exhibit defined and constant characteristics. Furthermore, any uncertainty surrounding the classification must become covered so as to maintain the solidity of the construction. The fact has to be rendered into an unquestionable reality in itself (Fleck, 1979). A good example of this erasure of uncertainty can be found in Martin and Lynch's (2009) description of the early research that was aimed at counting the number of human chromosomes. They found that agreeing on a number of chromosomes was far more important to the disciplinary community than actually getting the number 'correct'. This was because uncertainty surrounding the number impeded research and theorization. In the case of pandemics, ambiguity surrounding the level of threat leads to uncertainty in the level of risk posed.

In this way a key function of the pandemic phases was to reduce uncertainty. However, this function was not fulfilled. This was due to the fundamental ambiguity and contestation surrounding the phases. During the early stages of the threat, most of the critical discussion of the WHO's actions centred on its timing of the phase declarations. From the earliest discussions, the WHO's representatives offered suggestions that H1N1 was likely to result in a pandemic. For example, by only 26 April 2009 (immediately subsequent to the detection of H1N1), it was suggested that

> we appear to be in a situation where one of the swine viruses appears to be affecting [a] significant [number] of people in at least a couple of different countries in different locations. This situation has raised questions about whether we are entering into a pandemic period.
> (Fukuda, 26/04/09)

This provides some indication of the short timeframe during which the WHO was making its management decisions. From the initial development of the threat there was a clear pre-emptive expectation within the organization that H1N1 was likely to constitute a pandemic. Nevertheless, at the time, despite the suggested evidence of geographical spread (upon which the phases are defined) and the presumption that a pandemic was imminent, movement through the phases was slower than the phase guidelines would dictate. The WHO was criticized because the situation had often fulfilled the definitions of a certain phase long before the phase change was implemented. This was a result of the conflict between the function of the phases (to recommend action and measure risk) and their definition (which relied upon geographical spread). This discordance between the function and the actual measurement was the source of the failure of the phases as an effective classificatory scheme.

To fully explicate this failure, it is necessary to elaborate upon the way in which the WHO conceptualized each phase, and in doing so defended its decision not to escalate phases. At the time of the first alert, the H1N1 threat was set at a Phase 3 (limited human transmission) warning. The discussion early on revolved around what an increase to Phase 4 would imply and why an update was not taken more swiftly by the WHO. With regard to these issues, the WHO's representative suggested:

> If we go to Phase 4 because of the swine influenza virus, it basically means that we believe that the virus has significance, or a potential pandemic virus... that it is able to transmit from person to person and cause large outbreaks. This is a pretty big change from Phase 3 to Phase 4.
>
> (Fukuda, 26/04/09)

Again, note that even within these early stages of the threat, the potential for a pandemic was emphasized, but the increases in phase were not made by the WHO. The following is a description of Phase 4:

> In pandemic level 4 – 4 is basically when human-to-human transmission, or sustained human-to-human transmission is limited to one relatively contained geographical area and it is felt that there is a possibility of a containment effort being successful... [elaborates on containment strategies] ... So that is pandemic level 4, and basically as you can understand it is more of a very punctual concentrated local effort. And that is normally undertaken when there is only sustained human-to-human transmission in one given area.
>
> (Härtl, 27/04/09)

Given this definition, it is understandable that questions arose regarding the lack of phase increases. From the first few days of the threat it was clear that sustained human-to-human transmission in Mexico was

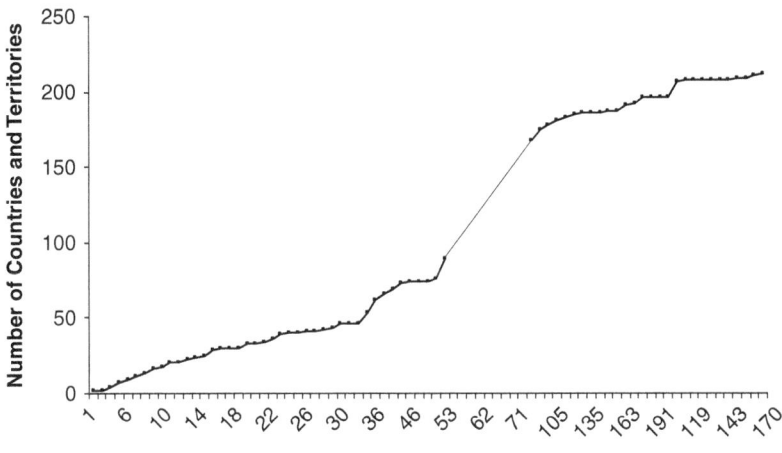

Graph 4.1 H1N1 geographic spread over countries and territories

indeed occurring, which fitted the definition of Phase 4 (WHO Situation Updates, 2009/2010). In fact, as shown in Graph 4.1, spread over multiple geographic regions (signifying that even higher phases could have been implemented than were announced) occurred very swiftly in the case of H1N1.

Graph 4.1, which has been produced using the WHO's daily epidemiological updates, shows that geographic spread of H1N1 occurred very quickly in the early period of the threat. (The gap indicates the period where data regarding spread across countries was not released). As can be seen here, within a few days the disease had spread across multiple countries. This created difficulty in the context of phases, the definition of which was based upon geography but sought to monitor risk.

Similar questioning of the WHO occurred when there was a lack of movement towards Phase 5. This phase represents an increase in potential impact and sustained community level outbreaks in at least two countries (WHO, 2009: 27), which, as Graph 4.1 shows, had occurred within a few days of the WHO's reporting of H1N1. At the time it was proposed that

> If we go to Phase 5 and 6, that basically indicated without going into details, that the virus has shown the ability and is spreading around the world. So it is really a geographical spread around the world.
>
> (Fukuda, 26/04/09)

78 *Pandemics, Science and Policy*

As this quote suggests, Phase 5 represents a large-scale mobilization against a disease. H1N1 had fulfilled the category characteristics very early on in events, but the ambiguity surrounding risk prevented the WHO from declaring the phase increase (reinforcing the inadequacy of the phase definitions). Again, the mobilization of preparatory actions indicated by this phase change is evident:

> So the question is in pandemic level 5, that would mean that we are seeing various foci of community transmission over several generations in geographically different places. If we make the determination, or if the determination is made now or in the coming days that there is sustained community level transmission, then that means that the virus is much more widespread and in that case there would be decisions needed to be made on switching over vaccine production from seasonal vaccine to pandemic vaccine. There would be other decisions needed on stock of antivirals... [and]... on health system provision because you can expect that if there are large, large numbers of cases then maybe health systems will be overrun so you would have to start triaging cases and ensuring that you deal with different types of cases differently.
>
> (Härtl, 27/04/09)

It is notable in these quotes that the concept of spread is emphasized rather than severity. With regard to moving to Phase 6 from here (a full pandemic declaration), spread is again noted as the chief characteristic:

> To go back to what Phase 6 mean[s]. The idea of Phase 6 was to capture how was this virus spreading and how far it has spread... we are really trying to get a handle on how far has it spread out and has it really established itself in different parts of the world.
>
> (Ben Embarek, 04/05/09)

Thus, the notion of severity is conspicuous here in its absence. In more clear examples of this, which serve to actually distance the phases from their purported purpose, it is suggested that

> There is some confusion about whether going to Phase 6 says anything about the severity of disease. These are separate issues and I hope everybody is very clear about it... Phase 6 means that we are seeing continued spread of the virus to countries outside of one region, and we are seeing community outbreaks occur in multiple regions of the world, it really tells us that the virus has established

itself, and that we can expect to see disease in most countries of the world. But that is different from the severity of the pandemic.

(Ben Embarek, 04/05/09)

And again,

When we look at the current Phases – the definition of Phase 6 – right now you see it is very clear. It simply says that you have community level transmission in a country outside of the region in which we are seeing transmission going on now at the community level.

(Fukuda, 26/05/09)

The exclusion of severity in the construction of the phases became the main site of controversy. The classifications depended upon geographical spread, but in the case of H1N1 such spread occurred very rapidly (as demonstrated in Graph 4.1). Thus the phases were designed to indicate risk but they depended instead upon spread – in H1N1 they faced a disease with high spread but (arguably) low global risk. The inadequacies in the definition of the phases (as indicators of pandemic risk) were thereby highlighted through this case, leaving the WHO and its phases susceptible to critique.

This lack of correlation between severity and phase changes rendered the classificatory scheme dysfunctional. The press questions (and the representatives' answers) with regard to the slowness of declaring phase increases illuminates some of the WHO's attempts to reconcile this discrepancy, and its inability to effectively do so. For example, early in the threat (26 April), one reporter questioned:

You said yesterday the Emergency Committee wanted to buy more time, but to many people here in the real world it looks very much like [this] does not fit the definition of Phase 3, and I am hearing from people in the infection control sector that this is really making life difficult for them. They don't know if they are going to 4, 5, 6, sometime in the next few days... Should they be operating under the assumption that this is probably a pandemic?

(Fukuda, 26/04/09)

To this the response was:

whenever we face any emergency situation there is of course a balance between the need to have a fair amount of information so that we feel that our decisions and our assessments are based on solid

80 *Pandemics, Science and Policy*

grounds. On the other hand, I think we are also mindful of the need for different groups to act and to make decisions. Now, if WHO goes ahead and makes a declaration that the phase has changed, then this is really a very serious signal to the world... We want to make sure we are on pretty good solid grounds.

(Fukuda, 26/04/09)

This exchange suggests that, in part, the (delay in) phase changes was motivated by the desire on the part of the WHO not to send 'strong signals' (which might instigate panic) until the decision could be adequately justified through evidence of impact as well as spread. It also reinforces the role of scientific uncertainty in the WHO's actions. In an additional early exchange, another reporter makes similar observations:

Given the description that you have just put out, it would seem that we have spread across continents, although there hasn't been community level transmission. Would it not have been expedient for the WHO to have perhaps stepped up some of its warnings or some of its actions given what we have now.

(Härtl, 27/04/09)

In response to this question the WHO's representative asserted:

I am not sure what WHO could have done that it hasn't already done. I think the extent of this event only became clear to the world in the course of the past week and already on the night from Thursday (23 April) to Friday (24 April) we went into 24 hour emergency mode...

(Härtl, 27/04/09)

Thus early in the development the WHO was already defensive, and heavily scrutinized on the matter of phases due to the fact that the progression of H1N1 did not conform to the expected progression of a pandemic as outlined by the phase definitions. The classificatory scheme reflected the organization's assumption that a widely spreading virus would necessarily signify high threat. However, this was not the case with respect to H1N1. Thus the phase classification did not fulfil its functional objectives, weakening the WHO's claim to authority in defining pandemic events.

Furthermore, the WHO did not wish to declare a pandemic or increase phases before member countries could be made cognizant of what such statements entailed or before the risk was more clearly defined. This demonstrates confusion in the practical implications of the phases. As examples of this, the WHO's representatives stated that

> Another question that has come up is that 'Are we in Phase 6 [full pandemic] now, or why haven't we declared Phased 6 now?' In here I think I simply want to say that we know that this virus is spreading and we are now seeing that activity is picking up in a number of countries and this is, as I had mentioned before last week, we know that we're getting closer to probably a pandemic situation. But in the period of time since we last discussed this, I want to you know that WHO has been working extremely hard in terms of preparing countries, in terms of preparing populations for what a potential move to Phase 6 or pandemic would entail.
>
> For example, there is work that is being undertaken right now so people understand really what does a pandemic mean, what does going to Phase 6 mean? Does this mean we are seeing something really severe change? Does this mean that there is a need for drastic actions to be taken? And here I point out that by going to Phase 6, what this would mean is that the spread of the virus has continued and that activity has become established in at least two regions of the world. It does not mean that the severity of the situation has increased and that people are getting seriously sick at higher number or higher rates than they are right now.
>
> (Fukuda, 09/06/09)

In these statements, although according to the classificatory criteria a pandemic already technically existed, the declaration was withheld until expectations of events could be better managed and until the WHO could more successfully define the case and position the global reaction. This represents a failure in the classificatory scheme, which set out to fulfil these definitional and performative functions but was actually ill defined.

The failure of the phases as a classificatory scheme is reinforced through an exchange where a journalist (David Brown, *Washington Post*) questions why the pandemic has not been declared since, by the WHO's own definitions, the situation appears to warrant it. In this exchange, Brown says:

> if you could please address the question of why there seems to be so much reluctance on going to Phase 6? It is a very clear definition.

The point was made, you know, long ago, that it does not measure severity. What is to be lost by saying that it is community spreading, in the community and more than one place – which it obviously is – more than one region, we are going to go to Phase 6 and it is a mild Phase 6. Why not just bite the bullet?

(Fukuda, 26/05/09)

In response was as follows:

The answer to that is really almost another question which is: 'what is to be gained by going to another Phase?'... Right now, when we look at the request: 'Why cannot WHO look at going to Phase 6' coming from the countries, there are a couple of concerns here. One of them is that in many of the countries they do not see H1 activity going on, and in these countries with the few cases, things are relatively mild. And so, behind that question is the sense that many countries are already doing things that are necessary right now to address the situation. But if you go and declare Phase 6 without very clear evidence that there is a sort of change in the global situation, it can lead to extra work for countries without much gain, it can lead to some level of panic, it can lead to some level of cynicism that something is being declared but which is not usefully producing something in terms of public health benefit and gain.

(Fukuda, 26/05/09)

This quote might serve to indicate that even at this stage (predeclaration) of the event the WHO was attempting to minimize potential future criticisms of prematurely/unnecessarily calling a pandemic and attempting to manage expectations, even though WHO guidelines suggested that the wide spread (and novelty) of H1N1 defined it as a pandemic strain almost immediately after its discovery. The WHO was put in this position due to the incongruity of the phase classifications with the general understanding of 'pandemic'. It needed to actively manage the member states' expectations in a way that the phases were actually supposed to have (seemingly 'scientifically') defined.

On the whole, the WHO's description of H1N1 as characterized by particular pandemic phases was unclear throughout the course of the pandemic. This was the result of the failure of the organization to effectively define the risk. The phases mirrored the geographical progress of a pandemic. However, the organization did not subscribe to its own definitions in declaring phases for H1N1. This was due to the fact that, while the phases are publicly assumed (i.e. by WHO member states) to characterize risk, they did not fulfil this function. In the case of H1N1, the phases were not escalated in accordance with their technical description due to the fact that the impact of the virus was indeterminate. However, this lack of definitional conviction led to questions regarding the validity of the phases and in turn undermined the legitimacy of the WHO in characterizing H1N1 as a pandemic.

Declaring the end of a pandemic: The problem of uncertainty

The inadequacy of the phase classifications was also highly evident in the later stages of events. The full Phase 6 pandemic was declared by the WHO on 11 June 2009. Almost immediately following this, towards the end of 2009, critics began to suggest that the end of the pandemic should be declared by the WHO. In particular, media questions centred on the suggestion that the decreasing spread and therefore caseload of infection indicated that the pandemic threat had passed. For the WHO, declaring the end of the pandemic was a difficult process for a number of reasons. In the case of H1N1, due to the fact that it had never been a severe pandemic, a clear 'end' was indeterminable. The Pandemic Alert Phases were ill defined in categorizing exactly what a post-pandemic period entailed. Eventually, following an extended period of ambiguity, the post-pandemic phase was declared by Chan on 10 August 2010, when she announced: 'We are now moving into the post-pandemic period. The new H1N1 virus has largely run its course' (Chan, 10/09/10). However, as with many aspects of the WHO's reaction, the declaration of a post-pandemic period was contested and characterized by a lack of solid definition.

Prior to the declaration of the post-pandemic period, the WHO needed to justify its continued assertions of the 'pandemic' (Phase 6) classification of H1N1. This was done through reference to the historical experience of pandemics. Similar to the allusions to the historical pandemic events outlined in Chapter 2, it was suggested that history should provide a guide for declaring the end of the pandemic. Thus

> in terms of how we move from a pandemic period to a non-pandemic period. Again if we look back at history for some guidance, we will see that we typically have a period in which pandemic infections are quite high. Then we go to a transition period in which those newly

84 *Pandemics, Science and Policy*

> emerged viruses, pandemic viruses, often become seasonal influenza viruses.
>
> (Fukuda, 03/12/09)

The question for the WHO, then, became one of distinguishing the point at which a transition period was entered (the pandemic period finished) and a non-pandemic period began. With regard to this distinction it was pointed out that

> the ending of the pandemic is not an on and off phenomenon, we really expect it to be more of a trailing off phenomenon, it does not happen overnight.
>
> (Fukuda, 18/02/10)

This illustrates the inadequacies of the classificatory scheme – it did not form a clear distinction between these two states (pandemic and non-pandemic). Furthermore, its effects were to portray pandemic events as clearly defined with a distinct endpoint. As a result, the WHO proposed that an ambiguous transition period was necessary, including

> the post-peak period which is the transition period as well as the post-pandemic period which signifies when we have quite a good expectation that we are really getting close to the normal period out of the pandemic period.
>
> (Fukuda, 18/02/10)

Here,

> The practical effect of indicating that we are in a post-peak period is really to give a broad signal to the world that even though we may continue to see pandemic activity that we expect that we are transitioning more towards a normal level.
>
> (Fukuda, 18/02/10)

However, although it was acknowledged that such a transition period was necessary, it was difficult for the WHO to effectively distinguish a point at which the end of the pandemic could be declared for H1N1, due to both its ineffective classifications and the mild manifestation of the disease.

For many months prior to declaring the end of the pandemic, the WHO's representatives had to constantly produce statements to justify the continued pandemic state. For instance,

Now we are about 8 months into the pandemic and one of the common questions coming to us is, that the pandemic is over, is it time to call it, and really the answer is, that it is still too early to make such a call.

(Fukuda, 17/12/09)

Now one of the first points reflecting questions coming to WHO is that it really probably remains too early to call the pandemic over.

(Fukuda, 17/12/09)

So again I want to point out at this time, we believe that it is too early to say that the pandemic is over.

(Fukuda, 17/12/09)

Now I think at this point again, I want to point out that the pandemic continues.

(Fukuda, 14/01/10)

In addition to such outright comments, the justification for the continuation of a pandemic was asserted in several ways.

The distinction with seasonal influenza presented an important point of note in the WHO's characterization of a pandemic (see Chapter 3). The end of the pandemic was likewise identified through reference to seasonal strains. Thus Chan suggested that, in defining a post-pandemic period,

First and foremost, we are looking for whether or not there would be out-of-season outbreaks, as we saw last year in both northern and southern hemispheres...

And further that a seasonal-like pattern needed to be observed:

Now, the second point is we notice that in countries with H1N1 transmission, the level of intensity is now moving back to a pattern similar to the seasonal influenza pattern. The third thing we observed in all these countries that we have been getting good data, there is no longer a dominance of the H1N1 virus as we saw last year. We are seeing a mixed virus pattern. By that we mean we see H1N1 virus; [but] we also see H3N2 and we also see Influenza B virus...

But, last... is that we are seeing some level of community-wide immunity, either due to natural infection by the H1N1 or due to passive immunity by vaccination.

(Chan, 10/09/10)

Paradoxically, in the case of H1N1, such distinctions were in fact difficult to make. Even the post-pandemic period disease was partially characterized by the WHO by the same features as the pandemic phase of the disease, due to the ill-defined nature of the categories and the lack of distinction between H1N1 and seasonal strains. For example, in comparing H1N1 with seasonal influenza in the post-pandemic phase,

> In addition, a small proportion of people infected during the pandemic, including young and healthy people, develop a severe form of primary viral pneumonia that is not typically seen during seasonal epidemics and is especially difficult and demanding to treat. It is not known whether this pattern will change during the post-pandemic period, further emphasising the need for vigilance.
>
> (Chan, 10/09/10)

The apparently 'defining' qualities of a pandemic were thereby rendered ambiguous in the accounts of a post-pandemic period, as they are apparently present in both contexts. This suggests that H1N1 presented a unique challenge to the integrity of the pandemic phases; the phases may not have undergone contestation, despite some weaknesses, had they not been tested by the simultaneous mildness and wide spread of the H1N1 virus.

The notion of future uncertainty was also presented in both the pandemic (see Chapter 4) and the post-pandemic phases, again blurring necessary distinctions between the two states. Here the WHO suggested that it was unclear what future course the pandemic virus would take. Thus, still within Phase 6,

> In terms of the impact of the pandemic, of the important point is that, from the very beginning of the pandemic, we had pointed out repeatedly that we don't really know what the future is going to bring. I think at this point, it is fair to say that we still haven't fully gone through the pandemic, and that it is possible that there could be unexpected events which occurs as we continue to go through.
>
> (Fukuda, 03/12/09)

Again mid-pandemic,

> in dealing with an event like the pandemic... there are fundamental limitations in terms of knowing what course the pandemic is going to take. We are always unsure whether there are going to be significant changes in the future but we know that that can happen and if they are going to happen we don't know what they might be or when they may occur. This is really something that we have always to deal with.
> (Fukuda, 24/02/10)

In this way, risk and uncertainty were highlighted during this period. Of these potential 'unexpected events' suggested by the WHO, possible future waves of the infectious spread were upheld as a distinct possibility and primary source of uncertainty in maintaining a pandemic state categorization. Thus

> One of the big questions which is still before us is whether we expect to see yet another wave of activity occur perhaps at late winter or in the early spring months and the answer right now is that we are simply not able to answer this question right now.
> (Fukuda, 17/12/10)

And also

> because it is unclear whether we will see in the northern hemisphere over the next few months during the winter and spring period another significant wave of activity and also because we do not know yet what will happen in the southern hemisphere during its winter months. So for these reasons, we consider that the pandemic is still ongoing.
> (Fukuda, 14/01/10)

As has been argued (Chapter 3), this state of uncertainty was emphasized as being characteristic of the pandemic event:

> Now from the very beginning WHO has gone out of its way to let everybody know that the future course of the pandemic was uncertain, that we did not have a crystal ball and could not tell you at the beginning, which way it was going to go.
> (Fukuda, 14/01/10)

In this way, uncertainty was often invoked in the WHO accounts, and this underpinned the continued labelling of H1N1 as a 'pandemic'.

However the WHO account also rendered the post-pandemic period as unpredictable. Again, this nullified attempts to distinguish the pandemic and post-pandemic states. Thus, in declaring the post-pandemic period, Chan suggested that

> Pandemics, like the viruses that cause them, are unpredictable. So is the immediate post-pandemic period. There will be many questions and we will have some clear answers for only some. Continued vigilance is extremely important and WHO has issued advice in recommended surveillance, vaccination and clinical management during the post-pandemic period.
>
> (Chan, 10/09/10)

To some extent an attempt to explain this discrepancy was made by the WHO through the suggestion that constant vigilance is necessary. However, this only reinforces the definitional ambiguity:

> As we enter the post-pandemic period, this does not mean that the H1N1 virus has gone away. Based on experience with past pandemics, we expect the H1N1 virus to take on the behaviour of a seasonal influenza virus and continue to circulate for some years to come.
>
> In the post-pandemic period, localized outbreaks of different magnitudes may show significant levels of H1N1 transmission.
>
> (Chan, 10/09/10)

Through these explanations the threat of H1N1 was not represented as significantly diminished in the post-pandemic state. This again indicates a failure by the WHO to distinguish the 'post-pandemic' from the 'pandemic', and a failure of the phases as a whole with regard to providing classificatory distinctions.

From the WHO's perspective, the experience of uncertainty may have underpinned its actions because it is easier for an institution to be overly cautious than to risk complacency, in that the consequences of non-action have potentially far greater repercussions than those for overreaction (Levidow, 2001; Stebbing, 2009). However, this lack of distinction between the pandemic and post-pandemic states within the phases rendered the WHO's construction of the event as a whole susceptible to interrogation.

Furthermore, during Phase 6, the WHO's representatives noted that the public and media calls for the declaration of the end of the pandemic did not take into account the global nature of its spread. From the WHO's perspective, the fact that the H1N1 virus had spread globally represented a vital characteristic in defining it as a pandemic threat. In this way the continued impact of H1N1 in specific regions served as justification for the continued pandemic classification. So,

> Based on the situation, our current assessment is that it remains too early to say that the pandemic is over. This is because we continue to see activity at elevated levels in a number of countries.
> (Fukuda, 14/01/10)

Thus 'pandemic activity is different at different places in the world', it was 'really too early to conclude that the pandemic was in a post-peak period in many countries' (Fukuda, 24/02/10) and for these reasons a post-peak period could not be declared at that time. Therefore

> I think that, if we look at how the world deals with these large global events... some of the recommendations made at the global level certainly are blunt because they are really intended to be relevant and germane to the world.
> (Fukuda, 24/02/10)

The global attribute of an influenza pandemic is highlighted in these justifications. Indeed, classificatory schemes often find problems when they attempt to consolidate localized and globalized problems into a uniform set of categorizations. There is often a disconnect between the locally experienced reality and the simplified global categorization (Bowker & Star, 1991; Mahajan, 2008). The WHO institutionally focused upon a global problem, whereas the member states experienced only national effects. To the extent that the phases served as signals for action on the part of member states, this was a fundamental inadequacy. This difficultly in consolidating the local and the global into one schema may have contributed to the overall inadequacy of the pandemic phases.

This lack of distinction led to significant consternation. Upon declaring the end of the pandemic, Frank Jordans (Associated Press) commented that 'Several countries started scaling back their H1N1 efforts some months ago, yet WHO held back on downgrading the pandemic phase until now. Why did it take so long?' (Chan, 10/09/10). To this the response was that

> Yes, indeed, what you said is correct. Many counties in the northern hemisphere in fact scaled back on their public health response to the

H1N1 virus... But having said that, the World Health Organization has a duty to monitor the global situation and that is precisely what we are doing...

Now all in all, we are seeing clear signals and evidence pointing to the fact that the world is now – and I'm talking about it at a global level – the world is transitioning out of the pandemic into the post-pandemic period.

(Chan, 10/09/10)

In a similar instance, the reporter Jules Caron asked that what he 'would like to know, between phase 6 and post-pandemic, what exactly does it mean? What is the WHO doing now that is does, didn't do before?' (Fukuda, 10/09/10). To this Fukuda responded that

this action simply notifies countries that we are transitioning out of a pandemic period in which we have seen unusual patterns related to influenza, back to a period in which we see influenza patterns more typical of seasonal influenza. However, during this period one of this things which we are strongly emphasising to countries is that it's important to continue monitoring and (stay) alert for unusual circumstances related to disease – this could indicate still ongoing severity of this virus – and also be on the watch for any changes in viruses.

(Fukuda, 10/09/10)

The evident need for vigilance in the post-pandemic period mirrored the WHO's discourse of uncertainty regarding the pandemic state. Here it is evident that there was a lack of distinction between actions occurring in the pandemic phase and those occurring in the post-pandemic phase – surveillance and control precautions were still emphasized. This again supports the argument surrounding the inadequacy of the phases in fulfilling their function of prescribing pandemic management techniques.

This is further evidenced by the fact that, despite the continuous assertion to the contrary, the notion that the pandemic was nearly at an end was also accepted by the WHO before the official post-pandemic announcement was made. This ambiguity in the organization's response reflected the ambiguity in the classification. Thus, prior to the post-pandemic declaration,

there may be some evidence that the highest levels are now past us. To be very succinct here, what we are hoping for is that the worst is behind us and that we are on a general decline in activity in some locations and we want to point out that even if we are entering into a period of general decline, we can anticipate that in some locations there could be significant local upsurges of activity.

(Fukuda, 11/02/10)

In this way the WHO attempted to simultaneously represent H1N1 as both a 'pandemic-in-progress' and as a 'pandemic-past' prior to the official declaration of the end of the pandemic, demonstrating the lack of sufficient classificatory demarcation.

Transition out of the pandemic state was always a matter of concern for the WHO. Alluding to the global nature of the declaration, and emphasizing the regulatory procedures surrounding such measures, it was suggested that the 'end of pandemic' was a planned-for occurrence:

The 2009 Pandemic Preparedness Guidelines anticipated at some point that there would be a transition out of a pandemic period but the world would not have reached a normal state in which we would be fully back into seasonal influenza patterns that we normally see in a non-pandemic period and this transitioning period where the pandemic activity continues but may be tailing down was really called the post-peak period. Again, the post-peak period can be considered a transitional period in which the pandemic is continuing but there is a scientific judgement that the worst, on a global level, is probably over – again even though there may be some local outbreaks occurring or local upsurges.

(Fukuda, 11/02/10)

Furthermore, the WHO represented itself as responding to expert opinion on this matter:

[From their referrals with 138 scientists from over 45 countries] they really indicated that the ending of a pandemic cannot be considered an abrupt on or off situation, that there would inevitably be a transition period...

(Fukuda, 11/02/10)

And

I anticipate that at least in some time in 2010 we will be discussing this in more formal settings, more concentrated ways, to try to get the best scientific picture, of where we are in this pandemic, whether we should expect a third wave in countries to come, or not, whether we think that this is convincing information to say that we are really moving away from the pandemic period.

(Fukuda, 03/12/10)

Thus, as with the distancing of responsibility in terms of creating the phases, the WHO also attempted to distance itself from the responsibility of declaring an end to the pandemic and, through the allusion to expertise, to suggest that the 'end' of the pandemic was an objectively definable event. However, despite these claims to procedure, it is clear that in reality the phases were vague and ill defined, so that the WHO had to retrospectively engage in definitions of the process as the events were occurring.

As with other aspects of the organization's depiction of the H1N1 threat, its narrative surrounding pandemic phases failed to effectively distinguish the concept of 'pandemic' and convey a sense of a genuine risk. In terms of defining the end of the pandemic, the WHO did not effectively depict the post-pandemic as distinct from the pandemic period – the classification was not precise. This rendered the notion of a pandemic, and the phases, susceptible to contestation. The WHO's attempts at retrospective definition merely highlighted the inadequacy of the initial phase definitions in categorizing pandemic disease, and they resulted in the legitimacy of the WHO itself being questioned.

(Re)defining phases: The problem of severity

To create a classificatory scale, as in the case of the pandemic phases, the phenomenon to be classified must be rendered measureable. That is, in order to classify the stages in a (potential) pandemic, it is necessary to be able to define measurements for a 'pandemic'. For something to be rendered measureable, it must contain an order to be measured (e.g. mild to severe, or pandemic potential to full pandemic), and it must contain a quantity that can be measured (e.g. geographical spread or incidence rate) (Swistak, 1990). Since the concept of pandemic was ill defined, it was also difficult to define a measurement. It is clear that in this sense the concept of 'pandemic' is a boundary object which is conceptualized differently by different communities of actors (Shackley & Wynne, 1996). During the timeframe of H1N1, the WHO defined a 'pandemic' by spread of disease, but this did not effectively correspond to other actors' conceptualizations of 'pandemic' (e.g. as a severe event). Thus the H1N1 'pandemic' is an example of a failed boundary object – the concept came under contestation.

Attempts to measure and classify boundary objects are necessarily tenuous. Choosing the order and quantity to measure is difficult because the entity is ill defined, and the choice may become almost arbitrary (simply in order to render the phenomenon measureable). In the case of the pandemic phases, the WHO's decision to focus upon geographical spread as a unit of measurement (rather than morbidity or mortality rates, for example) was a point of contestation following the case of H1N1 (see Chapter 7). The WHO needed to define the stages of a pandemic in order to erase uncertainty surrounding novel influenza strains, but the measure it chose did not fulfil the purported function of the phases.

As a result of its deficiencies, the WHO was petitioned by member states to reassess its phase-assessment criteria. This occurred in two significant events: the ASEAN+3 Summit (8 May 2009) and the World Health Assembly (WHA) (18–22 May 2009). At these events the representatives of the member states noted the confusion surrounding the phases and requested that the WHO should revise them.

With regard to these requests, the WHO said:

> But what did happen recently, at two large meetings – one of these was the ASEAN + 3 Meeting... as well as the World Health Assembly has requested WHO to look at the situation from going to Phase 5 to Phase 6, and to make sure that we are taking into consideration everything which ought to be considered... What that did was to really lead us to go back and reassess what are the needs of countries if we go from one Phase to another, and particularly from Phase 5 to Phase 6. What in fact is needed by countries to make that kind of movement helpful to them? This really did take us back to looking at severity, looking at the Phase criteria, and then consulting with a large number of experts and also public health staff for a number of different countries.
>
> (Fukuda, 02/06/09)

As these statements indicate, the criticisms made by the member states at the forums became an immediate point of discussion in the WHO's statements. Principally, the member states' confusion surrounded (the exclusion of) the concept of severity in the phases. The omission of severity as a defining classificatory characteristic led governments to question the legitimacy of the phase pandemic declarations.

They suggested that severity, rather than geographic spread, would be a more appropriate unit of measurement for the purposes of phase classification.

The criticisms revolved mainly around the inadequate measurement of risk, and it was noted by the WHO that,

> Really, two of the things that we are looking at in depth after the interventions from the countries is: what level of community spread really indicated that you have spread in the community. In addition, there are a lot of questions from countries about severity – does the impact on people make a difference in terms of going up to the [sic] Phase. These are two of the issues we are really looking at right now.
> (Fukuda, 26/05/09)

In this way,

> At the WHA [2009], what the countries raised was a concern and they said that currently the criteria from going to 5 and 6 are based on geographical spread, and this is true.
> (Fukuda, 22/05/09)

Again the problem of risk versus spread was evident. The member states understood risk as being synonymous with severity, while the WHO's phases measured only geographical spread:

> What has become clear is that it is not just the spread of the virus which is considered important by countries who really have to act upon the Phase changes, it is really the impact in the population. It is this input that has to be taken in and considered in terms of the Phase 5 to Phase 6 change.
> (Fukuda, 22/50/09)

> What it did was to reinforce to us that what countries are saying is that the spread of this virus is really a phenomenon that nobody can stop and that nobody can get in the way of, but in order to provide tools and guidance to countries, which is really helpful to countries, it is not enough just to say that we are at a certain Phase and that the virus has spread to a certain extent.
> (Fukuda, 02/06/09)

Such statements also reflect an attempt to deflect attention away from severity – again trying to emphasize spread as a legitimate measurement criterion. Such statements also somewhat misrepresent the member states' concerns, which appreciated that geographical spread did not necessarily relate to risk, and had not done so in the case of H1N1. This again illustrates the inadequacy of the phases with respect to helping member states to identify appropriate reactions. The inclusion of a measure of severity was seen as crucial for the countries because of the underlying assumption that increases in phases should correspond to increases in the predicted impact of the threat – this was the presumed function of the phases.

As the discussion regarding the phases and severity unfolded, differences in the perception and definition of these terms became increasingly clear:

> what the countries said is that we are in the mixed situation and we are concerned that if we go into Phase 6 the message to our populations will be: 'You should be very afraid', whereas in fact we [the WHO] think that it indicated that the virus is spreading out but the level of fear should not go up and there should not be an increase in anxiety.
>
> (Fukuda, 22/05/09)

Given the WHO's need to be responsive to the member states, the concerns were met by statements to the effect that the criteria would be reassessed, though they were still applied in relation to H1N1. The WHO made assurances that

> When we look at those issues and when we look at the complexities of severity, and the complexities of defining trigger points for moving up, then it seems like it is a reasonable thing to take stock, take a look at the situation and say 'really, what is the best way to proceed here.' It would be possible to simply say, well, because something is written down, we need to just follow those, that is the most important principle. But really if you take the perspective that the bottom line is what is it that we are going to do which is going to be helpful for people, which is going to be helpful for countries, then I think, hopefully, it puts it in more perspective of why we are looking at this so seriously.
>
> (Fukuda, 26/05/09)

This demonstrates that the organization had begun to realize the inadequacy of the phases. In addressing and alleviating these concerns, it was in fact suggested that continuous reassessment of actions constitutes an important facet of the WHO's organizational practice. Historical actions were again alluded to here:

> ...when we look at past situations that have been very difficult – and one of the most famous one was back in 1976 when we had the... [unintelligible]... swine influenza –... one of the overall big lessons, perhaps the single biggest lesson from that whole episode is: 'Take stock, take a look at what the reality is saying and do not put yourself in a hole and just leave yourself there'. You need to take stock of actions over and over again... we have a situation in which countries are saying: 'We want you to take a look at these criteria because if you apply them the wrong way, they may not help us. In fact, they may cause difficulties.'
>
> (Fukuda, 26/05/09)

The WHO thus agreed that the pandemic phase criteria needed to be reassessed. Due to the fact that severity was at the heart of the criticisms of the existing phase definitions, the concept became increasingly problematic at this stage in events.

This meant that the WHO's definitions needed to be defended. The omission of the concept of severity from the then-current phase definitions was argued by the WHO to be a result of attempts to clarify the concepts through simplification. Thus, when one reporter (*Science Magazine*) said:

> I am a bit confused. I think that WHO has always made it clear that a pandemic could be mild and it did not have to be a devastating one. So why was not the whole issue of severity never integrated into an alert system?
>
> (Fukuda, 26/05/09)

The WHO responded not by addressing the key issue of severity but by pointing to the complexity of phase definitions:

> The Phases themselves as planning tools have been around for quite a long time... Much of the feedback, when we were going through the revision process [between versions of the *Pandemic Planning Guidance*] was that the older pandemic Phases were too confusing. There were too many concepts in them, too many ideas in them and that they should be more straightforward and simpler to apply. The most recent version of the Pandemic Phases meet those criteria. They are much easier, they are simpler to understand, but... when you are

really addressing a real situation ... they probably do not adequately capture all of the concerns of countries.

(Fukuda, 26/05/09)

Thus attempts at simplification were put forward as the rationale behind the existing phase definitions. However, it was conceded that a more complex model was necessary to address real-world events. This reflects the difficulty in constructing classifications, which must strike a balance between simplicity and accuracy (Bowker & Star, 1991; Derksen, 2000; Freeman & Frisina, 2010). The WHO's 2009 phases were highly simplified but did not stand up to application with respect to the emergence of H1N1.

Immediately following these interchanges between the WHO and member states, the WHO resolved that the pandemic phase definitions would be re-evaluated in collaboration with member states and scientific experts. However, when the representatives later made references to discussions surrounding potential changes, the discourse shifted to an emphasis on the difficulties in including complex measurements. Thus it was stated:

> Now, over the last couple of days we have had a number of discussions related to Phases and to the severity of illness ... Yesterday at WHO, we held a series of consultations with a significant number of experts – over 30 experts and public health staff coming from 23 countries spread across the globe – and the reasons for these consultations was really to understand their perspectives and their concerns about a possible move from Phase 5 to Phase 6, and what considerations WHO should be mindful in doing so.
>
> These discussions were very fruitful for WHO and provided a lot of excellent advice and guidance, and there is also consensus in a number of areas. First, the experts advised WHO to continue using the geographical spread as a basis for moving to a pandemic Phase 6 with assessment of severity. In doing so, WHO should also provide more tailored guidance to countries, really to help them respond better to whatever the degree of severity of the situation is, in addition to just declaring that there is a Phase 6. Much of the discussion through the experts was over the matter of severity: how one makes such assessments with suggestions coming from the experts ranging from clinical assessments of illness up into economical impact and very large social measures.

(Fukuda, 02/06/09)

The consensus appeared to be that the WHO ought to include some indication of severity in its guidance to member states, but that severity as a concept cannot be integrated into the pandemic phases themselves. However, this would leave the situation as it was, in that a pandemic can technically be brought into existence regardless of impact, or it would be argued as being high risk. This demonstrates the problematic nature of 'risk' and 'severity' with regard to defining and identifying pandemic threats.

The WHO argued that severity could not be included as a characterizing aspect of a pandemic and phase changes because the concept itself is too complex. It asserted that

> Severity is one of those terms and concepts that mean different things to different people. What is severe to politicians, is different to what is severe to epidemiologists or what is severe to clinicians. The answer to that is that there is no simple clear epidemiological definition of severity. Here, in the way that we are talking about it, looking at it, is that our primary goal is to reduce the impact of disease on people and to reduce the adverse effects of that disease on people. The adverse effects are both directly the disease itself but also other aspects of it... But when people become infected, they may go to hospitals, may develop severe outcomes; this is a little bit easier to compare; so at least in terms of doing comparisons of severity between countries, one of the things we are focussing on first, is really the disease aspects.
>
> (Fukuda, 11/05/09)

Such reasoning served to justify the neglect of severity in the phase criteria. Indeed, as argued previously, severity itself can be regarded as an ill-defined boundary object. Since severity was too complex to measure, the WHO suggested that it could not be included in defining phases. However, this does not solve the problem because the function of the phases remains unfulfilled.

In further attempts the concept of severity became highly problematic and increasingly convoluted in definition. This is evident in the following attempts to explain it:

> when we talk about severity, it can mean different things to different people. For example, there is definitely clinical severity...
>
> There is also severity at social level and national level, in addition to personal level. But capturing this is really a very difficult

activity to do. How do you capture severity so that it is relevant for all countries, at the same time? This is a very difficult concept to capture. Nonetheless, the interventions from countries... should be taken into consideration... to see what kinds of adjustments might be made to make sure that the definitions really meet the situation.

(Fukuda, 26/05/09)

And again:

What is important to understand is that severity is basically based on three components:

The first component is the virus itself: its virulence and its transmissibility. But... we [also] need to take into account the vulnerability of this population... This is information we do not have at the moments because it is very difficult to assess the immunity of a population.

(Fukuda, 13/05/09)

The other factor – the vulnerability of the population – is also a pre-existing conditions... This third factor that is really important to understand the impact of a disease in a society, is what we call the capacity of the society to fight against this disease, or what we call also, the 'resilience' of this community.

So you see, because severity is not one factor... it is very hard to have an index, especially at [the] global level.

(Fukuda, 13/05/09)

In short, the WHO argued that severity is a concept which is so complex as to justify its redundancy – it was latterly depicted as a concept that is impossible to measure accurately. Thus, in reference to phases, the WHO attempted to discard the concept of severity altogether.

This of course contradicts the member states' requirement that severity should be included in any definition of pandemic phases. As a result, the WHO's final narrative of including severity was ambiguous. On the one hand there was a clear attempt to provide some measurements of severity in order to satisfy these criticisms. So:

One of the things that we will continue working on developing ways to assess severity and finalize these measures as soon as we can. One of the other things will be to provide more tailored guidance to help Member States so that they can better calibrate the actions. One of the things that we hope to do by providing this kind of tailored guidance is really to help reduce some of the more drastic actions, which may be uncalled for, but also to provide guidance to countries as to what steps they can take.

(Fukuda, 02/06/09)

However, it was also acknowledged that severity is difficult to ascertain, such that

> severity itself is assessed by other means, with the gathering of more detailed data, because it is not as straightforward as it is for example for the geographical spread. And there are also lots of questions about: will WHO issue a kind of severity index for the... global event? In fact this is an issue that has been discussed on many occasions in technical consultations before we issued the latest pandemic preparedness guidance, because it was really concerning. The assessment of severity is a key part of the information that will help national governments to plan for their response. Therefore, it is an issue that we have talked a lot about and we have also discussed whether it was feasible to have an index, at [the] global level...
>
> (Fukuda, 13/05/09)

In this way, and simultaneously, the inclusion of severity in the WHO's phase definitions was circumvented:

> what we have decided to do is not so much 'redefine' Phase 6, but to stay with the current criteria, really to augment the information provided when an announcement is made to Phase 6. Augmenting it really means coming out and explaining what we consider to be the severity of the pandemic, and also to come out with information for countries in terms of how to tailor [to] them some of [the] responses to the pandemic situation, which may differ from the pre-existing national plans.
>
> (Fukuda, 09/06/09)

In some cases, this reluctance was asserted as more open suggestions of the impossibility of including such measurements, as the WHO again attempted to side-step the issue:

> In fact for influenza, this kind of index is not very helpful, especially at [the] global level because severity will vary from place to place.

What we have seen in previous pandemics, and even in the same country, is that you can have different levels of severity. And throughout the pandemic itself, in previous pandemics, you have different waves and each wave can have its own level of severity. Therefore, to have one indicator to describe all this variety of situations was not very helpful.

(Fukuda, 13/05/09)

This latter viewpoint, that severity is 'unhelpful' and indeterminate, was a dominant narrative at the closing stages of the pandemic. Reviews of the pandemic phases are still under way to this point. However, sociologically speaking, the controversy surrounding phase categorization highlights the definitional ambiguity and serves to emphasize the overall failure of the WHO to successfully construct H1N1 as a true pandemic threat.

The Pandemic Alert Phases were an important element of the WHO's reaction to pandemic influenza threats. They were set as an indicator to member states of the level of risk and necessary action, but fundamentally they failed in this function. The phases failed to effectively communicate risk because they are based upon geographical spread rather than severity. Furthermore, due to the problematization of severity in the case of H1N1, the link between increasing phases and effective reactions became weak. The WHO's weak characterization of pandemic phases therefore led to the contestation of the phases, of the H1N1 pandemic and of the WHO's role in pandemic management. The problem of classification was integral to understanding the WHO's reaction to H1N1. As suggested in the sociology of classification literature, classificatory schemes are central to scientific knowledge production. The WHO's ineffectual classificatory device, the Pandemic Alert Phases, weakened its overall construction and management of the H1N1 virus.

5
Vaccines, Institutions and Pandemic Management

After a risk is constructed, a solution must be presented. A critical part of the WHO's institutional management of the H1N1 pandemic risk was its enactment of preventative strategies. It emphasized the importance of vaccinations in minimizing the impact of the pandemic. By referring to their historical utility and efficacy, vaccines were represented by the WHO as fundamental to prevention. Furthermore, the safety of vaccines and the role of manufacturers were constantly justified. Other potential reactions (including border control and antiviral use) were not similarly promoted by the WHO. This emphasis on vaccines ultimately became a pivotal point of criticism of the organization from its member states. The WHO's narratives of preparation were therefore central to both the institution's overall construction of the threat and the subsequent contestation of its position.

The analysis of the WHO's reliance on vaccines as a preventative strategy can be explained from a number of sociological perspectives. Perhaps the most developed of these is the political economic argument which suggests that the use of vaccines is fundamentally a result of the capitalist structuring of medicine. Such arguments persuasively show that networks of non-governmental organizations (NGOs), national governmental and corporations engage in profit-making interventions in multiple areas of global public health (Elling, 1981; King, 2002; Silverman, 1976; Vance & Millington, 1986). This Marxist political economy approach asserts both that such interventions are the outcome of a global capitalist structure and also that these networks serve to manufacture not only drugs but also diseases (Moynihan, 2002; Moynihan et al., 2002; Williams et al., 2008). The potentially 'manufactured' nature of the H1N1 pandemic could therefore be an important application of this approach, particularly with regard to pharmaceutical corporations.

However, the WHO would not have benefited from a false scare because this would, in fact, have served to undermine its credibility. Indeed the criticism of the Council of Europe shows the effect that such accusations can have, suggesting that the WHO would be wary of the potential for such allegations of collusion.

Following the focus of this book, the examination will remain upon the act of scientific fact-making within the institutional processes of the WHO. Of interest in this chapter is not necessarily 'who profits' from the use of vaccines but, rather, the mechanisms behind the clear institutional fixation upon vaccination as the preventative strategy, despite other possibilities (some which, such as antivirals, would be equally profitable to pharmaceutical manufacturers). The focus of this chapter is therefore on explaining the WHO's reliance and praise of vaccination as an effective pandemic-management strategy. It shows that the organization's emphasis upon vaccines reflects a continuation of its historical preoccupation with vaccination as an effective tool against infectious disease. Furthermore, the WHO's key historical successes came as a result of mass vaccination campaigns, which over time reinforced the organizational decision-making that emphasized vaccine use. In this way, institutional processes were key to the WHO's preventative strategies.

As has been demonstrated in the discussion of co-productionism, contemporary responses to risk often result in a (sometimes fairly arbitrary) choice from among a plurality of management strategies. Importantly, this process of scientific fact-making surrounding H1N1 occurred not only under conditions of technical and scientific uncertainty but also under specific institutional circumstances. There is a tendency for policy-makers to simplify and perceive problems in ways that limit the perceived scope of potential solutions (Janes & Corbett, 2009). Given the inherent scientific uncertainty embedded within the problem of H1N1, these institutional structures heavily determined the course of action. In particular, certain aspects of the WHO's reaction, especially its reliance on vaccination as a containment strategy, demonstrate that historical and bureaucratic institutional forces played a critical part in framing its response to the pandemic. As such, the institutional and knowledge-production forces interrelate within the H1N1 actor network in producing risk-management strategies.

In explaining the WHO's reaction to H1N1, it is important to elucidate the way in which institutions make decisions based upon historical practices (Douglas, 1989). The 'new institutionalism', characterized by a set of explicit (multi)theoretical accounts in the explanation of institutions, provides a useful basis for explaining this problem (Blyth,

2003; Hall & Taylor, 1996; Hay & Wincott, 1998; Quah, 2007; Schmidt, 2010). Importantly, the new institutionalism also broadened the definition of 'institution' to include informal norms/conventions/procedures (Lowndes, 2002, 2010). This contrasts with the former use of the term which referred narrowly to formal political organizations.

Although there is a renewed interest in institutions, the question of institutional origin and, more importantly for the present study, institutional change remains somewhat underanalysed (Pierson, 2000). The dominant approach tends to utilize basic functionalist reasoning, suggesting that institutions originate and subsequently change as a response to varying needs (Lowndes, 2010), or that they change endogenously through cumulative effects or minor shifts (Djelic, 2010; Mahoney & Thelen, 2010). In contrast, the potential for exogenous shocks to rupture and restructure institutionalized positions presents an alternative explanation (Djelic, 2010; Hall, 2010). Alternatively, change is understood as the result of the mobilization of power groups, either upon or within the institution, or the function of internal agents constructing new discourses that result in change (Hall & Taylor, 1996; Schmidt, 2010). However, one useful and well-theorized concept in explaining institutional change (or the lack thereof) is the notion of path dependency. As with the new institutionalism in general, the concept of path dependency has been utilized and modified by different theoretical paradigms. The most important distinction is between discursive and historical path dependency.

As the term implies, the concept of discursive path dependency suggests that the discourses produced by a given institution or between institutional actors frame actions. This approach is useful in analysing how certain core institutional ideas/narratives (e.g. liberalism and conservatism) impact upon decision-making processes (Blyth, 2007; Schmidt, 2008, 2010). A discursive path dependency approach suggests that institutional discourses construct and frame both responses and procedures. Here it is the discourse that creates actions, through the reinforcement and justification of the roles and goals of an organization. Furthermore, this approach also emphasizes decision-making as an outcome of the discursive processes that underlies it – the institutional actions vary according to the way in which the discourse surrounding a problem is constructed (Blyth, 2007; Schmidt, 2010). However, with regard to the present study, this approach is less useful given that the WHO's justifications of its risk management were framed in terms of scientific imperative and regulatory procedure. More generally, the broad discursive stances of the WHO (most evident in its founding constitutional declarations of commitment to equality of health outcomes and the more recent Alma Ata slogan of 'health for all') can be regarded as practically vague and somewhat divorced from the organization's routine bureaucratic workings (Beigbeder, 1998; Corrigan, 1979; Fee et al., 2008). While the organizational discourses that are evident in this chapter may in part strengthen the WHO's perspective on the problem, this discursive frame is itself a result of a historically reinforced structuring of institutional processes.

The historical institutionalist concept of institutional path dependency is more useful in explaining the WHO's reaction to H1N1. The concept of institutional path dependency suggests that historical context and historically conditioned decision-making tend to influence current and future institutional actions. It is particularly relevant to the study of the WHO's management of H1N1 because there are significant parallels with the organization's responses to previous infectious disease threats. While historical and discursive path dependency may be mutually reinforcing once the situation has been constructed, this case demonstrates that the historical context of the WHO was fundamental to framing its action with respect to H1N1.

Crucially, the concept of historical path dependency suggests that the current and future actions of organizations are likely to be influenced by historically contingent decisions and processes (David, 1994; Lowndes, 2010; Mahoney, 2000). The very nature of institutional processes renders organizations susceptible to such path dependency. Organizations are useful because they can acquire and process large amounts of information in order to achieve their function. The key to organizational processes is that information must be filtered, coordinated and simplified in order to be useful in decision-making (David, 1994). To do this, organizations develop information-processing procedures through which data are rendered useful for decision-making. Very often, these processes are determined in accordance with expectations made at the time of the organization's creation (Mahoney, 2000). Although processes can change slowly over time, the repeated use of a particular procedural 'code' is self-reinforcing. The organization tends to collect more information in the direction of its existing process, simultaneously becoming less efficient at acquiring and processing information that does not correlate with the existing procedural structure and outlook (David, 1994). This conceptualization of institutions coincides with other sociological theorizations about the self-perpetuating character of institutional processes – for instance, in the classic example of Weber's ([1913]1978) conception of bureaucracy, as well as Mary Douglas' (1989) argument surrounding the potentially irrational and deterministic consequences of institutional processes in *How Institutions Think*. Institutions are self-reinforcing, and (over time) the bureaucratic procedures are not necessarily rational in their outcomes.

In fact, the stability of organizational structures lies, in part, within this consistency of procedures. However, the same ability to coordinate action and provide a stable structure works as a barrier to change (Greenwald, 2008). For this reason the arguments of institutional path dependency carry significant explanatory weight. Due to this tendency towards maintaining stability, the historical context in which an organization is established essentially moulds the present actions. Thus the historically contingent conditions under which an institution formed 'can result in the selection of a particular solution for what was then perceived at the time to be the crucial generic function [of the organization]...' (David, 1994: 214). The organizational structure can become locked into a set of routines and actions, so that the present action is dependent upon past procedural pathways, despite the fact that the initial rationale may have become irrelevant. Even where change occurs, organizations typically evolve in a way that is shaped by preceding functions (David, 1994). This type of path dependency is evident in the case of the WHO's response to H1N1. Specifically, the WHO's past experience with vaccination strategies, which had been integral to the organization's early success, resulted in similar methods being applied to the H1N1 pandemic.

The utility and efficacy of vaccines

Under conditions of scientific uncertainty, governing institutions must make management decisions, despite the lack of evidence surrounding the problem. Historically, the WHO turned to mass immunization campaigns as a reaction to global public health disasters. It is therefore unsurprising that immunization was put forward as an effective strategy against H1N1, despite (or in fact given) the underevidenced nature of the problem. The use of vaccines was promoted by the WHO as an effective tool to combat H1N1. As such, the development of vaccines (in the early stages of the pandemic) and their use and distribution (following successful development) were prominent narratives throughout WHO documents.

It was suggested that 'all countries will need access to vaccines' (Fukuda, 24/09/09) to effectively deal with H1N1. From the initial discovery of the viral spread, vaccinations were focused upon as a valuable reaction. As the WHO put it,

Why are we so interested in vaccines against this new virus? It is because we all know that vaccines are an extremely effective public health tool and in addition, vaccines against seasonal influenza are protective against the disease – in severe disease – of millions of people every year. So, therefore, it is generally recognized and accepted that it would be critically important to have a vaccine if you want to stop the pandemic that might be coming with this virus.

(Fukuda, 01/05/09)

Vaccinations were therefore strongly advocated as the most effective method of minimizing the risk of H1N1 from the WHO's perspective on the pandemic.

The WHO provided evidence for this support of vaccines in a number of ways. Importantly, historical events were referenced as evidence of the effectiveness of vaccination against infectious disease. Theoretically, this reflects the important effect of institutional history in the reaction. Thus

It is clear when you look at the Twentieth Century that vaccines have been one of the most effective and most cost-effective and safest ways to protect people against a wide range of infectious diseases. Again, these include diseases such as yellow fever, polio, measles, meningitis, small pox, and so on. There is a list which goes on and on, but the idea is basically the same. It was true and is also true for pandemic influenza.

(Fukuda, 03/12/09)

As such the

WHO, along with other public health authorities believe that these vaccines are very useful against pandemic infection and do support their use. The second point is that these vaccines have now been used in a significant number of countries, vaccination programmes have started in over 20 countries over the past several weeks and, based on this experience in which millions of people have now received vaccines, we in fact see that these vaccines are very safe.

(Fukuda, 05/11/09)

In addition to arguing that vaccines are useful, this second quote reflects another primary concern of the WHO, which was to assure member states and publics that the use of vaccines is safe (a claim which will be addressed in further detail). The utility and dependence on vaccines was thus presented as unproblematic, and it was taken for granted that they would provide the most effective control measure. However, research surrounding the management of influenza pandemics suggests that other health measures (particularly social distancing and prophylactic anti-viral medication) are likely to significantly affect the impact of pandemic spread and associated morbidity – and potentially with greater efficacy than vaccination (see, e.g., Ferguson et al., 2005; Ferguson et al., 2006). The use of vaccines in the particular case of H1N1 was also criticized (see Chapter 7). The WHO's dependence on vaccines therefore reflects sociopolitical aspects of its institutional decision-making as much as it does scientific knowledge surrounding infectious disease.

This unquestioned dependence on vaccines can be understood as a result of the organization's historical successes with vaccination campaigns. While the WHO's official mandate ranges over all matters of health and illness, since its founding, it has considered the management of infectious disease as a key function. The first action taken in the early period of the organization was against malaria, tuberculosis and venereal disease, only secondarily targeting basic health services (Beigbeder, 1998). Over time, this focus upon infectious disease remained (Beigbeder, 1998; Fee et al., 2008). Although the WHO began to emphasize horizontal health programmes (e.g. integrated public health programmes and sanitation within each member state), its history has highlighted vertical disease campaigns. These have almost always been those against infectious disease (most notably smallpox, tuberculosis and malaria). Beigbeder (1998: 126) argues that for the organization,

> In contrast with the sometimes vague objectives of some of the WHO's programmes, the advantage of a vertical programme is to have identified a specific enemy, to know the technical means to control or eradicate it (immunization, effective medicaments etc.) and to be able to measure or evaluate the result of the campaign.

Campaigns against infectious disease have thus characterized the WHO's overall function and marked its key successes (and failures).

In particular, the WHO's campaign against smallpox serves as a prototypical example of its success in controlling infectious disease, and this is perceived by the organization as a critical historical juncture (Fee et al., 2008; WHO, 2007). The mass immunization campaign waged against smallpox provided a perspective through which the WHO has managed subsequent cases. Importantly, following this early success, vaccination became its dominant strategy in controlling infectious disease. These vertical campaigns against infectious disease have become prominent in the WHO's history and they were highlighted given that the success of such campaigns 'enhances durably and visibly the Organization's prestige. Inversely the failure of a programme is a public failure for the Organization, which may result in a loss of credibility and a loss of resources' (Beigbeder, 1998: 126). In fact, even the prominent failures of these campaigns lie in part in the WHO's dependency on vaccination as a resource against communicable disease. Most notably, the failure in malaria eradication was underpinned by the emphasis on finding a vaccine rather than relying upon other options (Beigbeder, 1998). This reliance was despite the fact that, to date, there have been no effective vaccines produced against human parasitic infection (Abath et al., 1998; Beverley, 2002; Da'dara & Harn, 2005). The reliance upon immunization represents a well-worn institutional response. This has occurred despite multiple examples where mass immunization was not, scientifically speaking, implicitly the most appropriate (and never the only possible) response. The early success with smallpox eradication, therefore, established a pattern of management which appears to be ingrained in the WHO's reaction to infectious disease threats; in short, a path-dependent reaction.

Thus, from the early stages of the H1N1 threat, the importance of vaccines as a control measure were emphasized by the WHO. Evidence of the historical efficacy of vaccination use was proposed as the rationale behind the WHO's focus on them in the case of H1N1. Given this emphasis on vaccination, in addition to issues of efficacy, the manufacturing methods and safety of such widespread immunization became an important component of the WHO's narrative of preparation.

The manufacture of vaccines

Due to the institutional focus on the use of vaccines, during the early period of the H1N1 threat, the process of manufacture was prominent within the WHO's overall discourse of utility. The capacity for vaccine production became a focal point of interest. From the WHO's perspective, the usefulness of vaccines was such that distribution needed to be global. In this regard it was questioned whether enough vaccines could be produced and (if not) whether equitable distribution was possible. As such it became necessary for the spokespeople to reiterate statements regarding the capacity to produce vaccines, suggesting,

In terms of capacity, there is much greater vaccine capacity than there was a few years ago. But there is not enough vaccine capacity to instantly make vaccine for the entire world's population for influenza.

(Fukuda, 05/05/09)

In more detail it was stated:

as you know we announced that the current capacity to make seasonal vaccine is around 900 million doses per year, and therefore, our conservative estimate is that this would translate to between at least one to two billion doses of H1N1 pandemic vaccine, if it should be a pandemic.

(Fukuda, 06/05/09)

In this way a large-scale effort at vaccine manufacture was planned from the beginning of the threat, although complete global coverage was deemed to be impossible.

An important part of the WHO's narrative of preparation (and projected role) thus lay in the just distribution and effective management of resources. Ethical distribution and use were emphasized. It was suggested that 'It is absolutely essential that countries do not squander these precious resources through poorly targeted measures' (Chan, 11/06/09). The question of appropriately managing vaccine resources was prominent in the WHO's account of the crisis. This provides further evidence of the organization's support of the use of vaccines and its depiction of its management role within global public health.

In addition to questions of capacity and distribution, the matter of timing of production was also highlighted. Given the large scale of the proposed production, the questions of when it would be necessary to approach manufacturers and how long it would take to develop the vaccines came to the forefront:

There are two important issues about the production of a new vaccine against this virus. The first one is when do you begin production of this new vaccine. Right now we are in a period in which there is work going on to develop a new vaccine. That is going on. It started almost immediately and that will continue to push on.

(Fukuda, 05/05/09)

Thus the question was never 'Should vaccines be used' but, rather, 'have we begun to produce them fast enough?' The production time was given as the reason why manufacture began almost immediately after the discovery of the spread, but the effectiveness of the strategy was only addressed in the broadest way.

Given the emphasis placed upon vaccines as a pre-emptive strategy, another considerable point of focus was the role of vaccine manufacturers. For the most part the WHO's representatives attempted to portray manufacturers as collaborative partners in public health campaigns. In part, such a framing was institutionally expedient, given the interrelations with multiple stakeholders that were necessary in the contemporary management of global risks. The WHO's main focus in its account of discussions with vaccine manufacturers emphasized their positive role in negotiating access/distribution to developing countries. In this way,

> one of the things that we have tried to really get across is that from WHO's perspective to make sure that some vaccine is made available to the developing countries is a priority for WHO and these are some aspects of the discussion that we are holding with these companies.
> (Fukuda, 22/05/09)

To attempt to ensure equitable distribution, there was an appeal made by the WHO to corporate ethics. The use of such claims can be seen in the following examples:

> We are organizing, as I speak, a meeting that has been called by the Director-General in Geneva on 19 May with the heads of all the companies making influenza vaccines... This will be a high level discussion with the manufacturers, appealing to corporate responsibility and to working together towards the increase of equitable access.
> (Fukuda, 06/05/09)

In addition:

> In terms of countries as I said, we are discussing with manufacturers, they are all aware of their corporate responsibility, they want to help WHO as much as they can in view of the already existing contractual agreement to provide access to WHO to this vaccine... after we have discussed and tried to secure as much as possible with the manufacturer, the other discussion will have to be placed at the political level between WHO and governments, to see how this can be played out.
> (Fukuda, 06/05/09)

As the quote above reflects, the WHO had to mediate relations between diverse actors in order to successfully manage global health threats. The nature of global risk management necessarily means that multiple parties must cooperate in the process. In particular, the nature of contemporary global health (see Chapter 8) results in greater influence for corporate actors in health policy. One of the most important features of global health is that all actors (states, NGOs and pharmaceutical corporations) are understood as 'partners' (Janes & Corbett, 2009; Maguire & Hardy, 2006; Reich, 2000). The private sector plays an increasingly important role in global health governance. However, this reliance upon the role of vaccines manufacture later opened up the WHO to critique regarding complicity.

Integrating the perspectives of multiple actors towards the goals of the organization presented considerable difficulties. For example, despite the institutional ideal of equality, ensuring vaccine access to developing countries is difficult under pandemic conditions due to prior contracts signed by (affluent) national governments. The WHO's depiction of the helpful pharmaceutical corporations grated in this regard:

> In terms of real time access, yes, this is what we are trying to sort out with the manufacturers. We are well aware, and they are not hiding the fact that they have agreements with a number of governments to provide access to vaccine. Most of the time, the contract will say a number of doses per week or per month. We are discussing with the manufacturers where they are in terms of filling up their books. And to make sure that in what is still remaining as available, that we would have access not to vaccines in six months, but some vaccines will be accessible already in the early weeks, and months of the production.
>
> (Fukuda, 06/05/09)

In reality, however, the vaccine manufacturers were a significant impediment to equal access:

> Of course the availability will depend on the manufacturers, dependent on the type of the agreement that they have with countries already, but we know and we have discussed with the manufacturers and most of them at least still have some window of opportunity in their orders. We want to make sure that we do not wait until that window has completely closed, and this is why we are taking a step now already before even having had a recommendation to go full scale to try to ensure access for developing countries.
>
> (Fukuda, 06/05/09)

Access to vaccines was therefore highly dependent upon WHO negotiations and pre-existing contracts. However, generally, the WHO attempted to portray vaccine manufacturers as associates in the management of H1N1 and partners in the public health effort.

Conversely, while much of the narrative depicts vaccine manufacturers in a positive light, at several points this narrative of mutual cooperation is contradicted by the portrayal of vaccine manufacturers as a profit-making enterprise in the more traditionally Marxist sense. For example, vaccines were shown to be a pure commodity in the sense of negotiations surrounding payment:

> Who pays? Of course there is always a question of money and there is a transaction cost. For the time being the manufacturers that have discussed with us have always either been very open to donation, I can remind you that prior to the H1N1, WHO has had donations from two companies of 50 and 60 million doses of H5N1 vaccine. There are companies that are still considering... donation to WHO for the benefit of developing countries, and also tier pricing. This is something that is really the norm, I would say, in the distribution of vaccine for poor countries, is that poor counties pay much less [for] their doses of vaccine than rich counties. Apart from that, who will pay? This again needs to be discussed, it could be donor countries, it could be charity, it could be development banks, and all will be put together to contribute to putting money forward.
>
> (Fukuda, 06/05/09)

Even more telling is the sense that the WHO has very little control over the manufacture of vaccines. For example, at one point it was suggested that, while the WHO's director-general could 'recommend to manufacturers to produce large scale stock of influenza A(H1N1)', ultimately 'This decision is not that of WHO as you know. The decision will be of the manufacturers to take. It is their prerogative to decide what they produce' (Fukuda, 06/05/09). Thus while the WHO can make suggestions about what vaccines might be necessary (and thus also profitable), there is no direct input upon whether manufacturers choose to make the vaccine. In total,

> a vaccine is like any other, in certain aspects, is like any other commodity that is produced by a manufacturer, a producer, and which is then marketed and sold to a private or a public customer. Now of course it is not completely like that because as we know there are a

lot of issues about public health, about equitable access and it cannot be considered simply as any other public good.

(Fukuda, 06/05/09)

This slightly liminal positioning of vaccines both as a public health good and a commodity sets up the ambiguous relations between the WHO and the pharmaceutical manufacturers, which are evidenced in this dual portrayal. Moreover, it helps to explain the WHO's pains to maintain congenial relations with the corporations. As mentioned in the introduction to this chapter, this more explicitly profit-driven aspect of pharmaceuticals will not be discussed in depth here (though the conditions of global public health will be elaborated on in Chapter 8). However, it is important to note here that these relations helped to frame the WHO's use and narrative surrounding vaccinations and the controversy that ensued.

While the WHO's dominant portrayal of vaccine manufacturers was positive, scepticism from outside actors was clear across the documents examined. For example, one press conference included several questions about the status of vaccine manufacturers, such as the following (John Zaracostas, press):

> what assurances do you have on the production capacity, from the industry, the big companies, the independent and the government owned entities?
>
> (Fukuda, 06/05/09)

In reply the WHO's representative spoke with enthusiasm about the responsible nature of the pharmaceutical corporations, but in a way which reflected the ambiguous relationship:

> How do we know (and I know that this is the question that you keep asking me) that actually what the manufacturers say is true reality and if they do have this capacity?
>
> First, what we know, is what they deliver on seasonal vaccine. And we know that they have produced for sure around 500 million doses of seasonal vaccine in 2009. So this we know for sure... Although WHO has absolutely no capacity and no mandate to go and verify production plants ('show that you can make so many doses... [unintelligible]..., if this is the case, or you can make so many eggs per week')... we really trust what the manufacturers tell us is the real truth. In addition, we know and they know that this is not the time to play games. They have always been very responsible

and we do trust that what we say in terms of capacity represent what is currently available.

(Fukuda, 06/05/09)

This optimism regarding the actions of manufacturers foreshadowed the later controversy surrounding the use of vaccines, as is evident in the accounts of the Council of Europe (Chapter 7).

While the WHO concerned itself with questions about manufacture and just distribution, the question of whether mass vaccination was the ideal method in the first place was not addressed. Due to previous experiences with such campaigns, the WHO took for granted the idea that vaccination would be the most efficacious strategy against H1N1, demonstrating the path-dependent nature of decision-making. Due to the reliance on pharmaceutical companies, the narrative surrounding such corporations was necessarily congenial. However, this confidence was somewhat eroded when external actors started to question both the efficacy and the safety of the H1N1 vaccines.

The safety of vaccination

The presence of institutional path dependency is further evident in the WHO's narratives which served to reinforce suggestions of the safety of mass vaccination. However, following the criticisms made at the Council of Europe, the safety of vaccines became a highly contested topic. Despite the WHO's reassurances, one of the most potentially damaging critiques of the organization was the suggestion that the use of vaccines was actually unsafe. The WHO's documents that have been analysed often directly addressed these concerns. However, throughout, the reliance on vaccines never wavered. In combating criticisms, statements such as the following were released:

> One of the most basic questions to ask about vaccines is why are these being promoted? Why are these useful? I think here the answer is relatively straightforward and simple. We are in a situation in which the world is seeing a new infection, this pandemic influenza. This is an infection which clearly can cause death or serious illness in a number of people... We now have vaccines which are developed specifically against this infectious disease... we now have good evidence based on many people receiving the vaccine, but have no picture of unusual side effects emerging... So the picture right now looks quite good in terms of safety.
>
> (Fukuda, 05/11/09)

Such statements served to both continue reinforcing the suggestions of the utility of vaccines and provide assurance regarding the safety of these measures. Generally speaking, there are a few prominent narrative features through which the safety of vaccines was reiterated by the WHO. These were the suggestion that institutional regulation guarantees safety; the suggestion that widespread use has proved the safety of vaccines; and the comparison of H1N1 vaccines with seasonal influenza vaccinations.

The WHO's representatives were at pains to illustrate the regulatory measures that had been put in place to monitor the production of vaccines. The organization pointed to its own institutional mechanisms of regulation as the manner in which safety could be guaranteed. In this way, institutional process again shaped the discourse surrounding preparations, here with the suggestion that effective regulation produced an effective response. Through the production of institutional discourse suggesting effective regulatory practices, the uncertainty inherent within risk management was described as minimized. This mirrors many contemporary institutional efforts at risk management (Levidow, 2001; Marshall & Picou, 2008; Ravetz, 2004; Rothstein, 2006; Rothstein et al., 2006; Saloranta, 2001).

Importantly, H1N1 resulted in the first widespread use of non-traditional vaccine manufacture methods. This became an early point of concern. From the perspective of the WHO, these new vaccine technologies represented an important and heavily regulated advance in pandemic control measures. Thus:

> What about new technologies? They look very, very promising. There is no doubt that in the future...we will have other vaccines that will not be made in eggs. But we don't know how large the production will be in a small amount of time. The difficulty at this stage is that at the maximum these candidate vaccines have been tested in what is called Phase 1 clinical trials.
>
> (Fukuda, 06/05/09)

Where

> This Phase 1 clinical trial is a few doses in healthy adults usually. There is a big leap of faith to say that a few doses in individuals have been vaccinated and that you can take this very same product and inject it in millions of people. This is why for all novel vaccines, fantastic innovations, some of them are really great, but all these innovations need to be tested very thoroughly in clinical trials

and the dossiers reviewed by National Regulatory Authorities before authorization to deploy them is given for large scale implementation at the present.

(Fukuda, 06/05/09)

This is because there is

the obligatory step will be to test it in humans, in clinical trials, and following this, national regulatory authorities will need to register their approval on this vaccine before it can be made available to all. This is absolutely mandatory and obligatory because by no means do we want to compromise the safety of otherwise healthy people by inoculating them with a product that would not have all the guarantees of safety and quality.

(Fukuda, 01/05/09)

In this way, regulation became a matter of focus, and it was indicative of safety from the perspective of the WHO. In total, the WHO's accounts suggested that

What we would like to avoid is to say, we don't know whether it will be safe or not be safe, so let's gamble on safety. I don't think WHO nor any regulatory authority wants to gamble on this. You may remember that there were some difficulties in 1976 with the last scale mass vaccination campaign in the US against an outbreak of swine flu and nobody would like to repeat this experiment.

(Fukuda, 06/05/09)

For the WHO, then, the manufacture (and safety) of vaccines was a matter of interest early in the development of H1N1, foreshadowing the later focus and debate on the subject. The allusion to the 1976 Guillain-Barre episode became the focus of critical comment at later dates but was used in this earlier context to emphasize the interest in regulation and safety. Here the WHO used the example to explicate its concerns around the strict regulation of vaccine production.

In accounting for the safety of vaccines, the procedural aspects of vaccine developments and the regulatory mechanisms for safety control were emphasized by the WHO. For example,

there have been concerns voiced in the press [that] mainly the time lines for the developments of these vaccines is so quick that it would not assure safety. So I would like to make it clear that for all vaccines have a safety profile [to ensure safety].

(Kieny, 06/08/09)

The concern over the timelines was voiced predominately with respect to the non-traditional lines of vaccine. It was suggested by the press and critics that these vaccines were developed too quickly (within a matter of months) for their safety to be guaranteed. In this case also the WHO's representatives pointed to the sufficiency of the regulation methods:

> Now for pandemic use there are number of vaccines which are not the classical seasonal vaccine [i.e. non-traditional methods]... [which the WHO] would test extensively in clinical trials... [as] this is a registration of a prototype.
>
> (Kieny, 06/08/09)

During the manufacturing stage

> the data and all the controls on these vaccine lots are being submitted right now to regulatory agencies to look at the data presented by the manufacturers, and to take decision on their safety and suitability for use in the population.
>
> (Kieny, 06/08/09)

In this way, reference to regulatory measures represents one of the strategies through which the concerns about the safety of vaccines were assured in the WHO's representation.

Another method through which vaccine safety was demonstrated by the WHO, primarily in the later stages of events, was by reference to feedback received from the release of the vaccines (known as market testing). The argument here was that the use of the vaccine without any record of adverse reactions provided proof of the vaccine's safety. Thus it was asserted that 'In terms of vaccination, we estimate that over 150 million doses of vaccine have been distributed in about 40 or more countries.' (Fukuda, 03/12/09) Similarly, it was argued:

> we have now seen that over 300 million people, an estimated 300 million people or more, have now been vaccinated against pandemic influenza and that the safety record of the vaccine has been excellent. We have not seen any unusual safety events occur.
>
> (Fukuda, 24/02/10)

The fact that vaccines had been used successfully was therefore cited as evidence for their safety.

However, the most prominent argument put forward by the spokespeople was in attempts to link the H1N1 vaccines with those against seasonal influenza. Specifically, it was suggested that the tested use of seasonal flu vaccines suggests that the H1N1 vaccines are equally safe. Thus Fukuda stated: 'I would also like to point out that part of the vaccines are based upon very old and proven technology which are used for seasonal vaccination.' He continued: 'No new safety issues have been identified from reports received to date... reporting so far reconfirms that the pandemic flu vaccine is as safe as seasonal flu vaccines' (Fukuda in Kieny, 19/11/09). This comparison with seasonal influenza vaccines served to assure safety because, 'for seasonal vaccines, millions upon millions of doses have been administered to all kinds of populations, including very young children and including pregnant women' (Kieny, 06/09/09). As such, 'so far we have not seen any unexpected safely issues emerge, and the safety profile continues to be similar to what we see in seasonal vaccines' (Fukuda, 03/12/09). Thus, while in all other cases the WHO attempted to distinguish H1N1 from seasonal influenza (see Chapter 3), with regard to vaccine safety these links were continuously emphasized.

The topic of vaccine safety was pivotal in the wider debate surrounding H1N1 and the WHO's handling of the situation. Thus, as had been noted by some reporters, 'vaccines are undergoing a lot of scrutiny, there is a lot of scepticism about them in some very affluent countries'. This suggests that the WHO could have been 'concerned that this mass vaccination campaign could actually create problems for the reputation of vaccines' (Kieny, 06/08/09). In response to suggestions that the controversy surrounding H1N1 vaccines might cause wider distrust of vaccination, it was argued:

> We hope not because vaccines are really one of [the] prevention methods against infectious disease which is best in terms of efficacy, the safest and really we are always worried when there are rumours of vaccine safety and most times, these rumours are unfounded so it needs to be reacted to very quickly.
>
> (Fukuda, 06/12/09)

As such, assuring vaccine safety, and dispelling the rumours surrounding vaccines, was vital to the WHO's agenda. Consequently, adverse reactions and in particular cases criticism and popular reaction against vaccination were particularly focused upon throughout the WHO's texts.

Adverse reactions

It is clear that the WHO reacted to H1N1 through the use of vaccinations in a largely uncritical way. Vaccinations were considered to be an effective strategy because they had proved to be so in the past. Thus the organization was reacting to H1N1 through the procedural lens of their historical management of communicable disease. However, this path-dependent view of vaccinations was quickly problematized by outside criticism of its actions. Specifically, the safety of vaccines was questioned in such a way as to cause the WHO's representatives to reflect upon the choice to react to H1N1 with a mass vaccination campaign.

One of the more prominent points of controversy surrounding the use of pandemic vaccines was the suggestion that they were liable to cause Guillain-Barre disease. This suggestion had been made by a number of commentators and, most notably, was a key allegation made to the Council of Europe by Wolfgang Wodarg and associates (refer to Chapter 7). Guillain-Barre syndrome is an autoimmune disease affecting the nervous system and occurs when an immune response mounted against a foreign antigen misrecognizes and attacks host nerve cells. While most commonly manifesting as a complication of the immune reaction against a bacterial or viral infection, Guillain-Barre can also be produced as a result of immunization (Hughes et al., 1999).

In 1976, following an outbreak of swine flu (H1N1) in the USA, a mass vaccination campaign was mounted by the US Government. Subsequently, several hundred people who had received the immunization became affected by Guillain-Barre Syndrome (Evans et al., 2009; Haber et al., 2004; Langmuir, 1979). As such, although the cause of this rise in incidence has not been definitively established, Guillain-Barre Syndrome has been linked with swine flu vaccinations since this event.

Critics of the WHO's use of mass vaccination in response to the 2009 H1N1 outbreak drew links with the 1976 case to suggest that the vaccines were unsafe. This argument was reinforced by critics' suggestions that the organization's recourse to vaccination had occurred as a result of the influence of the pharmaceutical industry. In response, the WHO noted that there had not been a sizable increase in Guillain-Barre cases as a result of the 2009/2010 H1N1 vaccination campaign, suggesting that

> There has been in particular a lot of concern about Guillain-Barre syndrome because of the incidences during the swine flu vaccination campaigns in 1976 in the US. To date, less than a dozen suspected cases of Guillain-Barre have been reported following [2009/2010]

vaccination. Only a few of these Guillain-Barre cases may be linked to the pandemic vaccine. Illness has been transient and patients have recovered.

(Kieny, 19/11/09)

Given the large numbers (millions) of individuals vaccinated in the 2009/2010 campaign, the WHO suggested that the rate of Guillain-Barre should be considered as coincidental; these incidences did not represent a repeat of the events of 1976. Furthermore, it was suggested that advances in manufacturing methods represented an important point of difference between the two cases:

> The vaccines [of the 2009 and 1976 cases] are very different...the degree of purity that is obtained now with the vaccine which is used now for seasonal [flu] and... of pandemic vaccination was much less advanced in 1976 so there were many more impurities in the vaccines...So the vaccine that we have now is much purer and the quality controls and testing in the laboratories which is made of today's vaccine is much better that that of 30 years ago.
>
> (Kieny, 06/08/09)

It was thus argued that Guillain-Barre did not represent a realistic matter of interest in the case of H1N1 vaccination, and the WHO was clearly concerned to provide evidence against critical accounts which made such links.

The question of adverse events in general was often evident in media questions. However, as the WHO's spokespeople pointed out, 'Given the scale of vaccine administration, at least some rare adverse events could not be excluded' (Kieny, 19/11/09). In response to the somewhat repetitive lines of media questioning surrounding (specific and general) instances of adverse reactions, it was asserted that such cases were inevitably going to occur:

> Because when you vaccinate thousands and millions of people, of course, some of them were going to have heart attacks, some women are going to have miscarriages, and these may – because they have just been vaccinated – be associated in the minds of people with vaccinations. So there will need to be some careful analysis, to try to see which adverse event, which is really a worry, is associated with vaccination, and which are only co-incidental.
>
> (Fukuda, 24/09/09)

As this statement emphasized, adverse reactions can often be only a temporal coincidence – an illness which occurs following vaccination cannot necessarily be causally linked to the vaccination itself. However, these coincidences can drive media and public criticism. In another example the WHO argued:

> Well, of course the most serious of these events is death and there has been a small number of deaths. But again a report doesn't mean that this even is linked to the vaccine or that the vaccine is the cause. But they have been investigated so the more severe event, as I say, is death and of those we have heard around 30. Then potentially Guillain-Barre but Guillain-Barre we had a dozen of which only a few can be possibly due to the vaccination. And all these have resolved without sequelae.
>
> (Kieny, 19/11/09)

Mirroring the more specific example of Guillain-Barre, the repetitiveness of the press questions suggest that vaccine safety was at issue, and the WHO's responses attempted to placate these concerns surrounding adverse reactions of inoculation.

It is clear from the case study of H1N1, and given an understanding of the organization's history, that the WHO perceives mass vaccination campaigns to be central to its work. Accordingly, it was at pains to counteract any criticisms of the use of vaccines with regard to H1N1. The claims of critics, mobilized into popular movements (see Chapter 7), served to problematize the WHO's reaction to H1N1 and, furthermore, to problematize the organization's overall legitimacy. Such challenges came as a exogenous 'shock' to an institution (Campbell, 2010; Gorges, 2001; Greif & Laitin, 2004) in which vaccines had previously been conceived as unproblematic in the management of global infectious disease. The ensuing criticism of the WHO's vaccination strategy also rendered all actors more aware of the plurality of potential management strategies, highlighting the path-dependent nature of the WHO's reaction. The organization's defence of vaccines as useful because they had 'worked before', and had successfully undergone market testing, only served to demonstrate (to critics) that the WHO relied upon vaccines in an unreflective manner. Its reaction can be explained sociologically through the tendency of institutional processes to perpetuate certain types of decision-making and through the co-productionist argument that decisions made under conditions of scientific uncertainty are necessarily somewhat arbitrary (and are therefore susceptible to path dependency). However, the WHO's reaction to H1N1 served to bolster the general critique of its institutional processes.

Other preparatory actions

It has been demonstrated that vaccines were overwhelmingly considered by the WHO to be the best line of defence against H1N1. However, other public health measures were acknowledged as secondary measures at certain stages in the discussions. These include antivirals, border restriction/quarantine and surveillance/monitoring. The discussion of these measures is important both in highlighting the central role of vaccines, since other measures were only discussed fleetingly, and in illuminating the other potential prevention strategies which might have been considered useful had the WHO not been so institutionally focused upon vaccination. Some of these, such as border control, were indeed understood as primary prevention strategies by other actors, especially national governments. Overall, the WHO's narratives surrounding these measures merely served to illustrate the focus on vaccines.

Antivirals

From the initial development of the threat, it was clear that vaccines constituted the predominant reaction to H1N1. Thus, while antivirals are often the focus of many public health campaigns against influenza-like illness (Hayden, 2006; Lipsitch et al., 2007), in the case of H1N1 they were not a subject of focus at any stage throughout the threat. This was clearly illustrated in the suggestion that

> We do not really have a very strong position on the use of antivirals. It is part of the National Pandemic Preparedness Plan and, of course, we do not have any experience with treatment and clinical efficacy against this new virus.
>
> (Shindo, 12/05/09)

This quote strongly reinforces the argument that the WHO disregarded other strategies at an early stage. While antivirals were a part of the official planning procedures, their use was foreign. However, to some extent the potential efficacy of such strategies was acknowledged, for example, when it was asserted that [the anti-viral] oseltamivir can be a very useful intervention for treating people who are sick and so I think there is no reason to hold back from using it because we are concerned about resistance.

(Ben Embarek, 04/05/09)

It was also argued that

Part of this will be guidance that we are soon publishing, and we will recommend to consider the use of antivirals for high-risk groups of people at increased risk, depending on availability.

(Shindo, 12/05/09)

In this way, antivirals were not defined by the WHO as appropriate as a general preventative action in the case of H1N1. Rather, they were seen as a secondary measure, to be utilized for high-risk groups and severe cases. Thus while antivirals were supported for use against H1N1, this action was restricted to specific groups and not regarded as of major importance in combating H1N1 for the general public.

The WHO's narrative surrounding antivirals was vague and only present at the early stages of events. For example,

In the initial guidance, we took a more conservative approach because we had almost no experience with regard to the effectiveness of the antiviral medicine in this disease, and also we were aware that access to the influenza medicine was very limited. Now, we have gained knowledge in effectiveness, safety of the medicine and we have also contributed to the global availability of the medicine.

(Shindo, 12/11/09)

In this quote, again, the lack of familiarity with antivirals was stated. With regard to the 2009 H1N1 pandemic, it was argued that 'people in at-risk groups need to be treated with antivirals as soon as possible when they have flu symptoms' (Shindo, 12/11/09), but the WHO

want[s] to stress that people who are not from the at-risk group and who only have typical cold need not take antivirals. We are not recommending taking antivirals if otherwise-health people are experiencing only mild illness, or as a preventative measure in healthy people.

(Shindo, 12/11/09)

This dismissal of antivirals in the case of H1N1 (as opposed to the reliance upon antivirals during H5N1 and other threats) is an interesting issue. In part, this may be due to the notion of antiviral 'wastage' and mismanagement during previous disease scares (McCaw & McVernon, 2007). Notwithstanding the fact that H1N1 did not become resistant to antivirals (so that antivirals could have presented an effective measure throughout the pandemic), they were not emphasized by the WHO as a control measure and were rarely discussed in the texts. Despite the official acknowledgement of antiviral use, the institutional focus on vaccination was pivotal in the oversight of alternative strategies.

Border control and quarantine

Confusion and controversy surrounded the preventative strategy of border control and quarantine as a possible pre-emptive reactions against H1N1. Sociologically, collective understandings of infectious disease tend to produce narratives of prevention which reflect notions of threat, morality and blame (Bashford, 2002; King, 2002; Nelkin & Gilman, 1991). As infection is transmitted through social interaction, social distancing, isolation and quarantine are often a major part of collective responses towards infectious disease (Abeysinghe & White, 2011; Foege, 1991; Gensini, 2004; Herzlich & Pierret, 1987). However, the WHO's lack of interest in these problems and perceptions was evident in the examined documents. In terms of the restrictions on travel, it was made clear early on that 'the Director-General recommends not closing borders or restricting travel' (Härtl, 27/04/09). It was stated that

> [the] WHO does not recommend closing borders and does not recommend restriction of travel...with the virus being widespread, from the international perspective, either closing borders or restricting travel would really have very little effect, if any effect at all, stopping the movement of this virus.
>
> (Fukuda, 27/04/09b)

The organization suggested here that the already generalized nature of the spread suggested that actions such as border control and quarantine were ineffective. Again, the peculiar nature of H1N1, which was characterized as rapidly widespread, rendered common public health techniques such as border control unproductive. In another example it was suggested:

> One of the main considerations in Phase 4 is a potential effort to try and stop this virus, which is normally called 'containment' or

'rapid containment'... [but]... based on the analysis of the current situation and particularly because the virus is so widespread... that really this virus is too widespread to make containment a feasible consideration.

(Fukuda, 27/04/09b)

However, the WHO also contended that

given the current situation, the current focus of efforts should really be on mitigation efforts rather than trying to contain the spread of this virus. Predominately because this virus has already spread quite far, and at this time, containment is not a feasible operation.

(Härtl, 27/04/09)

In addition, it was reiterated:

Just to remind people, one of the decisions of the Director-General of WHO is that we do not recommend border closures and we do not recommend restrictions of travel... However, we are very focused on the safety of the people who may be infected with this new virus or with any other infection and so, with regards to that point, WHO does strongly recommend that people who are sick should strongly consider deferring travel.

(Fukuda, 28/04/09)

Thus, as this quote suggests, travel restrictions were not regarded as useful in the context of preventing the spread of H1N1. However, it was recommended that the sick did not travel in order to protect their own personal health:

From the international perspective when we look at whether travel advisories may slow the spread of infection, may slow the spread of the epidemic, we believe that at this time, these kinds of manoeuvres would not substantially help to do this, and so we are strongly emphasizing a focus on the safety of travellers...

(Fukuda, 28/04/09)

At the same time the WHO argued:

We do very much think that all steps should be taken to protect people. In terms of travel, again two of the most important pieces of advice are that if you are feeling ill, before you begin travel or before you begin air travel, you should strongly consider to delay that travel and stay at home until you're feeling better and not symptomatic... These steps will help ensure the safety of people who are getting ill. We will also not disrupt travellers, it will minimise the disruption to travel. This is the advice that we will provide at this point.

(Fukuda, 30/04/09)

In this way, border control and quarantine were not considered to be effective measures by the WHO. However, such statements did not stop some countries from instituting such measures. As a result there was some discussion surrounding the potential uses for these measures, despite the WHO's initial disregard of them.

Although the institutional focus was on vaccination, the WHO's narrative of prevention was also a response to the actions of its member states. As a result of questions and criticisms of the actions of several member states in terms of quarantine and border control, the utility of such actions was illustrated by the WHO's representatives. For example, with respect to member state actions it was asserted:

Just to talk in general about quarantine, I will remind everybody what quarantine is. Quarantine is when you have people who are not sick, who are not showing symptoms and they go into an area that is quarantined off, so you minimize the contact between them and other people. The instances in which this kind of control measure is taken is, if you're very early in the spread of a disease, you may use quarantine to try to limit the spread of the disease. That is one reason why you may institute quarantine.

Another reason you may institute it, is that you know people have been in close contact with someone who is sick... So there are different reasons when you can apply quarantine. It really depends on an assessment of what is going on and what you are trying to do at that time. So it is not a simple yes or no – you should do it, you shouldn't do it – you need to analyse the situation and then make your decisions about whether to apply quarantine.

(Fukuda, 05/05/09)

However, as always, the WHO's spokespeople refrained from commenting directly upon the actions of individual countries, given that the organization perceives its roles as a provider of (essentially non-directive) information (see Chapter 8). Thus they argued:

> Let me talk a little about disease control to put this into perspective. When you are dealing with infectious diseases, quarantine has been a long established principle... It is a bit different to isolation. Isolations is when somebody is sick and you put them to the side a bit so that they reduce the chances of infecting others. I do not want to comment on the specific disease control actions of different countries. I do want to point out that quarantine, in specific situations, can be applied and it is a quite reasonable action to take in specific situations. There are different times when it would be reasonable and other times when it would not be reasonable. In the guidelines pointed out, or developed by the World Health Organization in terms of pandemic Phases and preparedness, if you go to that document again you will see that there are considerations of when to apply quarantine, [and/or] when to apply isolations as considerations. But as we have mentioned over and over again, the situations differ and countries' approaches to disease control measures are choices. There is no set recipe of how you approach disease control and so this will differ to some extent from country to country. So I will leave it at that.
>
> (Ben Embarek, 04/05/09)

This quote emphasizes the WHO's ambiguity surrounding isolation, quarantine and border control. However, the key point was that, in any case, its discourse positioned national governments as ultimately responsible for these measures. It also emphasized:

> In terms of airport measures, disease control actions by different countries reflect the decisions based on considerations in that country. Over the past few weeks, I have not specifically said that I think that countries should do this or that or have not commented on the disease control actions taken by countries, but I have pointed out that there are a number of different actions that countries can take, and so leave it at that, but these are really country level choices.
>
> (Fukuda, 11/05/09)

As such, care was taken to avoid engaging directly with criticism of the actions of specific countries, and the question of border control was dealt with lightly.

Nevertheless, the WHO's statements served indirectly or privately to influence the action measures taken by member states. For example, it was suggested that, in terms of particular measures taken,

> This is under discussion with a number of different countries. I will not go into specifics of the countries, and again as I have mentioned before, I do not want to talk about any actions taken by any country.
>
> (Fukuda, 05/05/09)

However, it argued:

> We have talked about disease control again in a number of these press briefings and as you know from the guidance put out by WHO for different Phases, we have laid out the principles and the different disease control actions which can be considered by countries. And then these ought to be applied depending on the situation of the country and depending on the specific circumstances. So right now we are in a situation in which a number of different countries have instituted different kinds of disease control measures. One of the things that we are doing with these countries is contacting them to ask them about their actions.
>
> (Fukuda, 05/05/09)

In addition, it asserted, more forcefully:

> At the early stage when we met with the Emergency Committee, based on the evidence we made some recommendations. Clearly, no closure of border[s], no restriction of travel, and also no trade ban and we make those recommendations. Recommendations are recommendations, and we did see that some countries are not following the recommendations coming from WHO under the IHR. But under the IHR, I have a duty: require them to provide me with the public health justification on taking those actions.
>
> We keep chasing after all the countries and ask them to explain why they are doing what they were doing. And I am happy to say that things are getting better, but we must recognize that with a new disease, with a new threat, with a lot of uncertainty, it is not unusual to have a degree of overreaction and in some quarters they described it as panic. I think this is understandable, it is acceptable and we do need to give people the right kind of information to allow them to make that adjustment reaction. And we are seeing that this is being

done very well and the countries are lifting all these bans that they have imposed in [the] early phase.

(Chan, 11/06/09)

Thus the WHO claimed some jurisdiction over governments' actions through its characterization of the different control measures. Due to the rapid spread of H1N1 early on, it strove to minimize border control and quarantine measures because the disease, from its perspective, had already spread rapidly beyond control and vaccines provided the most effective solution.

Monitoring/surveillance

With respect to influenza pandemics, and particularly the WHO's global role, monitoring and constant surveillance are often asserted as prominent contemporary management techniques (Baker & Fidler, 2006; Declich & Carter, 1994; Martinez, 2000; Mykhalovskiy & Weir, 2006). With respect to H1N1, as with vaccines, monitoring was described as pivotal to the WHO's activities and essential to mounting a competent reaction against the threat. Thus the WHO emphasized:

> The first thing that we continue to stress with countries and continue ourselves to be very alert as to what the disease activity is. We are stressing monitoring as a very critical first activity, and as we keep saying the situation is evolving. The only way we are going to know how does it evolve and are there any important changes is by ongoing surveillance around the world and the participation of all countries.
>
> (Ben Embarek, 04/05/09)

In addition it argued:

> I think that at this point the most important thing we can do now, as it will be later, is to maintain a very high state of monitoring and watchfulness... this is really much of the emphasis on how we deal with these infectious diseases in the 21st century which is really to use every means possible to keep on top of this threat, monitor because we know that they can change very quickly, so this is what we are doing.
>
> (Fukuda, 28/04/09)

While there were relatively few explicit references to monitoring and surveillance in the examined documents, it is clear that the WHO saw these measures as critical to its overall functioning and role in the global public health system (Baker & Fidler, 2006; Declich & Carter, 1994; Martinez, 2000), particularly given the embedded uncertainty of the 'evolving' pandemic situation. As has been demonstrated through the discussion of the pandemic phases, the monitoring of epidemiology and spread represents a pivotal institutional function of the WHO.

However, overall, in terms of practical reactions to H1N1, it was clear that the WHO favoured the use of vaccination. Other common influenza control measures, such as antivirals and border control, were presented as inefficacious in the organization's account. Vaccines were presented as having been historically proven as an effective measure against infectious disease generally, and influenza specifically. Due to this reliance on vaccines, the organization presented a largely positive description of vaccine manufacturers. Furthermore, both the utility and the safety of vaccines were heavily defended. The emphasis on vaccines was later targeted in criticisms of the WHO's actions, as will be shown in Chapter 6.

The actions of the WHO provide a clear illustration of some of the potential negative consequences of organizational path dependency. Institutions tend towards stability and consistency, yet change can be vital to successful management (Greenwald, 2008). In the WHO's case, in order to react effectively to disease threats, adaptations must be made in reaction to both political and economic changes and epidemiological variation in their target diseases. However, the WHO was clearly path dependent to the extent that it persisted in pursuing one option (i.e. mass vaccination) over other possibilities (i.e. more traditional public health measures or antivirals), due to its successes with that strategy in its early history of communicable disease management. Nevertheless, contemporary influenza pandemic threats tend to be characterized by a rapidity of geographic movement and a fundamentally different aetiology, which distinguishes them from the past conditions which favoured mass immunization measures.

The WHO's own reliance upon allusions to its historical victories against infectious disease in justifying its 2009/2010 reaction lends further credibility to the suggestion that it was path dependent in its reactions to H1N1. These reactions have formed what Mahoney (2000) calls a 'self-reinforcing sequence'; there had been an initial formation and subsequent long-term reproduction of an institutional pattern surrounding infectious disease governance. However, organizations which find that they do not adapt effectively in reaction to changing contexts often face decline or dissolution. The Council of Europe's criticisms of the WHO's response to H1N1 in Chapter 6 highlights the erosion of the organization's perceived legitimacy. In this way the path-dependent reactions of the WHO in relation to H1N1 have proved to have a significant negative effect on wider perceptions of the its role within global public health.

6
Contestation and the Council of Europe

As this book demonstrates, the construction of a scientific 'fact', such as an H1N1 pandemic, is a product of multiple social forces and relations. In the context of scientific uncertainty, institutional decisions regarding risk management must occur despite a scarcity of evidence. This need to act upon the perceived threat, combined with the presence of a multiplicity of perspectives surrounding contemporary global risks, served to render the WHO's risk management actor network fragile and open to interpretation and critique. This chapter presents a case study of one prominent institutional challenge to the actions of the WHO in managing H1N1 – the critique mounted by the Council of Europe. Politically, the Council of Europe challenged the WHO's use of vaccines as a risk-management strategy. However, as this I argue, such a critique was only made possible through the contestation of fundamental aspects of the 'science' of H1N1. Sociologically, I demonstrate the fragility of the H1N1 actor network through an illustration of the Council of Europe's contestation. I furthermore demonstrate the democratized nature of contemporary science, where an outside actor – the Council of Europe – was able to impinge upon the WHO's internal institutional processes.

All aspects of the WHO's representation of the H1N1 pandemic threat were contested by the Council of Europe. In fact the WHO's management of H1N1 was (and at this time continues to be) a site of intense controversy. This is explained sociologically as the result of competing conceptualizations surrounding H1N1, which were a consequence of the WHO's failure to effectively bring about 'closure' and establish the H1N1 pandemic as a scientific fact. The Council of Europe mounted the most prominent and first organizational and political voice of criticism against the WHO. Aided by the benefit of hindsight, it emphasized the mildness of H1N1 in criticizing the WHO's management. The way in which the Council of Europe represented H1N1 therefore provides a telling contrast with the WHO's narrative of its management of the H1N1. This chapter, which explores the Council of Europe's account, will be structured following the themes of the previous substantive chapters of this book: the construction of the influenza and the H1N1 virus; the construction of 'pandemic'; the construction of risk; the declaration and definition of Pandemic Alert Phases; the use of vaccines as a risk-management strategy; and, finally, the Council of Europe's construction of the WHO's role in global public health. The juxtaposition between the accounts of the two organizations demonstrates both that the H1N1 threat could be differentially conceptualized, and that the WHO's construction of events was weak and ineffectual.

The Council of Europe's interest in the WHO's handling of H1N1 began at the end of 2009. One of the loudest voices of criticism came from the German epidemiologist/physician and Council of Europe parliamentarian Wolfgang Wodarg. He was the first institutional critic of the WHO's handling of H1N1, and he emphasized what he described as the undue influence of pharmaceutical manufacturers upon the WHO's actions. He presented a recommendation, endorsed by 13 other members, to the Council of Europe on 18 December 2009 entitled 'Faked Pandemics: A Threat to Public Health'. The motion suggested that

> In order to promote their patented drugs and vaccines against flu, pharmaceutical companies have influenced scientists and official agencies, responsible for public health standards, to alarm governments worldwide. They have made them squander tight health care resources for inefficient vaccine strategies and needlessly exposed millions of people's health to the risk of unknown side-effects of insufficiently tested vaccines.
>
> The 'bird flu'-campaign (2005/2006) combined with the 'swine-flu' campaign seem to have caused a great deal of damage not only to some vaccinated patients and to public health budgets but also to the credibility and accountability of important international health agencies.
>
> The definition of an alarming pandemic must not be under the influence of drug-sellers. The member states of the Council of Europe should ask for immediate investigations in the consequences at national as well as European levels.
>
> (Wodarg, 18/12/09)

This motion foreshadowed what would become key themes in the debate surrounding the actions of the WHO, namely assertions of the undue alarm caused by the declaration of a pandemic and the inappropriate influence of the vaccine-manufacturing industry upon the organization's actions. This was all associated with the primary claim that a 'true' H1N1 pandemic did not exist; the issue of definition is again prominent. The claims were investigated through several key discussions and committees of the Council of Europe, which form the basis of the analyses made in this chapter.

The assertions of Wodarg and his associates revolved around four main themes. These included the claims that

- H1N1 could not be considered a pandemic;
- the WHO caused undue panic in its handling of the case;
- this was due to the influence held by pharmaceutical corporations;
- the products of these manufacturers were not merely unnecessary and ineffective but also dangerous.

The Council of Europe's enquiries concentrated on an analysis of the role of the WHO in what it characterized as the costly and wasteful reactions to H1N1. At times the speakers and parliamentarians were highly critical of, and polemical against, the WHO, asserting conscious manipulation of the situation. For example, it was suggested that 'Everyone had been a victim of a chain of massive deceptions' (Diaz Tejera (representative for Spain) in Council of Europe Parliamentary Assembly, 24/06/10). On the whole, the mood of the proceedings is highlighted by the following:

> Our message is a powerful, thunderous and intelligent one of anger against a foolish act by the World Health Organization. We are the first body in the world to look at this problem and to denounce what happened. This is not going to go away.
>
> (Flynn (rapporteur) in Council of Europe Parliamentary Assembly, 24/06/10)

However, while these overtly political aspects of the proceedings are interesting in themselves, for the purpose of this book the focus will be maintained on using the Council of Europe's narratives as a case study to indicate the lack of conceptual closure surrounding H1N1 and the difficulty of managing risk where scientific evidence is indeterminate. This failure to reach closure thereby rendered the H1N1 actor network unstable at the most fundamental level.

Co-productionist analyses provide some indications of ways in which the contemporary structure of science means that scientific policy can become contested. For example, the democratization of science (the opening up of scientific institutions to public debate) provides a greater avenue for the criticism of science policy than in the past when scientific fact appeared to be more certain (Nowotny, 2003a). Under previous conditions there was a clearer distinction between insiders (scientists and scientific institutions) and outsiders (the rest of society). However, in the contemporary era, while some boundaries are maintained (and while continuous boundary work seeks to strengthen authority), 'outsiders' have far greater input and ability to critique scientific endeavours. This is because (due to the conditions of risk and expertise) scientific institutions are incapable of producing conclusive answers. This means that 'outsiders' have a greater ability to force themselves into the scientific dialogue. Where the debate about science is conducted before the public, such 'outsiders' may be able to criticize the scientific institutions and even set the agenda (Funtowicz & Ravetz, 1993). This can be seen in the incursion of the Council of Europe upon the knowledge-producing authority of the WHO. It is clear here that the WHO had lost full authority over the management of global public health, and the greater movement towards institutional transparency and 'democratization' forms part of the reason why criticism of the organization became possible.

The maintenance of boundaries of authority is pivotal to the acceptance of science and scientific policy, and the cost of failure to an institution is high. If boundaries between science/non-science, science/politics and experts/policy-makers are not maintained then knowledge and policy will be subject to contestation. In this case an allegation of a conflict of interest, and declarations that the science of H1N1 reflected the flawed processes of the WHO, reinforced the critique of the WHO's policy. The critique was made more possible in a climate which Wehling and Boschen (2004) refer to as the rise of a 'reflexive governance of knowledge' in the management of risks (Braun & Kropp, 2010). This suggests that there is a greater chance of debate and contestation of the production, regulation and application of the science that surrounds risks and (importantly in the present case) the ideas and institutions which conduct this management. The legitimacy of any policy decision rests upon the ability to reconstruct a plausible scientific rationale for the action (Jasanoff, 1987). However, the reflexive governance of knowledge means that these rationales are more likely to be publically scrutinized, and the tenuousness of scientific evidence that surrounds risks suggests that policy-making institutions are more easily subject to criticism. This chapter illustrates these arguments by directly comparing the Council of Europe's account with the themes drawn from the WHO's account.

The nature of the virus/the nature of influenza

When an actor network fails, multiple associated concepts (i.e. linked actor networks) can come under observation as black boxes are opened or destroyed. Within the actor networks of scientific institutions, this can then lead to contestation of what constitutes scientific 'fact'. As demonstrated earlier, within the WHO narrative, the understanding that H1N1 causes the disease of influenza was a taken-for-granted reality. However, this primary assumption was questioned by the Council of Europe's investigations. In fact, this fundamental divergence provides a good example of the way in which an apparently unquestionable scientific 'reality' (that H1N1 causes the disease influenza, and that influenza is a harmful disease) can become contested at points of scientific dispute, where closure had not been definitively established.

Contrary to the WHO, the Council of Europe depicted the concept of 'influenza' itself as problematic. This was first suggested by one of the key scientific experts who was called upon by the Council of Europe in the March meeting, Dr Tom Jefferson. Following this meeting, the ideas proposed by Jefferson were integrated into the official documents produced by the Council of Europe's committee. Pivotal to the argument was the suggestion that it is impossible to differentiate between influenza-like illness (ILI) and 'true' influenza. Thus it was argued that

> Influenza surveillance programmes in different places appear to report on the presence and degree of threat of influenza but what they are really looking at are influenza-like illness/flu.

And therefore

> we cannot say for certain how much influenza is circulating as influenza is an unknown proportion of an unknown whole (influenza-like illness/flu).
>
> (Jefferson, 29/03/10)

It was therefore suggested by the Council of Europe that the WHO, through its global influenza surveillance programme, made no distinction (or did not measure the distinction) between ILIs and influenza. Extending from this, it was maintained that much of what the WHO proposed to be H1N1 was actually not influenza at all.

According to Jefferson's account, the failure to distinguish between ILIs and influenza resulted in the WHO's misplaced reaction. Specifically, like most of the Council of Europe's narrative, this suggestion asserted the misuse of vaccinations. Thus

> vaccination programmes are directed against what surveillance systems worldwide call 'influenza' but in reality are influenza-like illness/flu. Surveillance systems cannot distinguish the two and provide reliable estimates of impact. This point is the key to understanding what comes next. The false equation 'influence-like illness/flu = influenza' has misled some of the research on the effects of influenza vaccines and (most of all) the interpretation of such evidence.
>
> (Jefferson, 29/03/10)

And again,

> Another consequence is the idea that influenza-like illness ('flu') and its ravages can be prevented or minimised with influenza vaccines... vaccines could only affect **at the most** (i.e. if they had 100% efficacy) some 7–15% of the annual flu burden, since this is the proportion of people with the flu who truly have influenza. This 'specificity' of approach (go for influenza, disregard all other cases of flu) is probably based on what I call availability creep... But, if you think about it, it is a wonderful utopian policy against a syndrome as unspecific as this (just think of the role that other viruses play). In my opinion, the lack of logic in this thinking is stunning.
>
> (Jefferson, 29/03/10)

The Council of Europe's critics' accounts thus suggested that the WHO mistargeted, such that

> the currently available evidence does not allow us to know in a reliable way how many cases of influenza there are, nor its impact in terms of death and disability with any degree of certainty. However, the confusion between influenza and influenza-like illness ('the flu') has led to an obsession with a single agent (the influenza virus) which is not based on any sound evidence and, as I hope you now realize, is potentially dangerous and misleading (because even a perfect vaccine can not work against influenza-like illness/flu as a whole).
>
> (Jefferson, 29/03/10)

In this way the problematization of the nature of H1N1 supported the Council of Europe's main point of contention elaborated below regarding the (mis)use of mass vaccination campaigns. Important here is that basic scientific assumptions can become questioned in the event of scientific dispute. Here, the notion of 'influenza' and its surveillance were deconstructed.

This depiction of the conflation between influenza and ILIs was taken up in the official documentation produced by the rapporteur Paul Flynn. For example, it was stated that mortality rates had been inflated due to this

> With regard to such a possible overstatement [of risk], the rapporteur would notably like to point out that, in many countries, no clear distinction had been made between patients dying *with* swine flu (i.e. showing symptoms of swine flu whilst having died of other pathologies) and patients dying *of* swine flu (i.e. swine flu being the main lethal cause).
>
> (Flynn, 23/03/10: 3)

From this perspective the threat of H1N1 had been magnified because the WHO had failed to take into account differences between illness that merely presented like influenza – ILIs – and 'true' influenza. What the WHO had stated to be 'swine flu' H1N1 was therefore, according to the Council of Europe, not necessarily influenza at all because surveillance systems were unable to effectively distinguish between different forms of respiratory illness.

In addition to the Council of Europe questioning the 'fact' of influenza, the veracity of the claim that H1N1 had pandemic potential was questioned. As has been shown, one of the main features that are characteristic of a potentially pandemic influenza strain was, according to WHO guidelines, the novelty of the viral agent. However, the Council of Europe's narrative contradicted the WHO's assertion that the 2009 H1N1 was a novel strain. Here,

> the WHO declared... that this was an entirely new virus. Now what here we see [sic] on the 22nd of May in 2009, we see that 10% of the under-60s and 30% of the over-60 age bracket already have an immunity against this virus. So we say, 'well, why stage things in this way, why manipulate things in this way?' when the virus is used in this way.
>
> (Rivasi, 29/03/10)

In this quote the presence of immunity in certain populations underpinned the assertion that H1N1 was not a novel strain of influenza, thereby asserting the argument that H1N1 was highly unlikely to cause a pandemic. In fact the events were said to be 'staged'. The viral threat was thereby deconstructed from an objective reality (in the WHO version) to an object that was manipulated ('used') in order to achieve political ends.

Wodarg and Keil suggested that influenza is typically a mild illness of little concern, and that the 2009 H1N1 strain in particular was indistinguishable from seasonal flu. Thus Wodarg claimed that H1N1 is a 'mild flu. People fall ill as they usually do in winter season' (Wodarg, 26/01/10), and furthermore that the extent of illness and especially severe respiratory symptoms associated with ILIs 'is considerably less than in previous years. Thus, not only is H1N1 not an unusual and novel threat but it is furthermore claimed that the incidence of illness is actually lesser than the typical influenza season' (Wodarg, 26/01/10; emphasis added). Keil's address reinforced the statement that H1N1 did not represent a novel threat, suggesting that 'the H1N1 virus is not a new virus, but has been known to us for decades' (Keil, 26/01/10).

The Council of Europe therefore put forward a fundamentally different account of the nature of influenza and H1N1 from that proposed by the WHO. Through its narrative it suggested that H1N1 wasn't novel, threatening or even distinguishable from seasonal influenza and ILIs. Thus the concept was contested at the most basic level of the nature of both H1N1 specifically and influenza generally, demonstrating the malleability of scientific 'fact' under conditions of dispute and uncertainty.

What/when is a pandemic?

Another major point of conceptual contestation is found in the definition of 'pandemic'. Within the Council of Europe' narrative, suspicion surrounding the WHO's declaration of the pandemic was prominent. It was maintained that the WHO's decision to declare H1N1 as a pandemic was erroneous. This assertion was reiterated through several key points of argument, many of which utilized the WHO's own 'evidence' to make the case. The question of what constitutes a pandemic was heavily disputed.

Epidemiological statistics, and the way in which the accounts of the WHO and the Council of Europe each employed them, were a recurring theme in the debate surrounding the validity of labelling H1N1 as a 'pandemic'. Wodarg and Keil pointed to epidemiological aspects of the H1N1 virus to suggest that in fact this particular virus should never have been recognized by the WHO as pandemic-causing. The morbidity and mortality statistics of the disease were cited as evidence of this proposition. For example, Wodarg said:

> Given the fact that the influenza is always a very contagious disease which spreads very rapidly and leads to a greater number of cases, it is surprising to see the extent to which attention was focused on that flu [H1N1] after the reporting of only hundreds of cases.
>
> (Wodarg, 26/01/10)

Furthermore, the epidemiology of the virus was suggested by Wodarg to be indicative of its non-threatening nature. He argued:

> Those who are over 60 years of age hardly contracted the [H1N1] flu. There is a relatively higher number of young people who contracted this flu which is not surprising at all. Usually, when we observe a flu coming, one of the factors, which helps us determine if it is already known or not is the occurrence amongst the elderly. If they do not fall ill they seem to already have immunity...
>
> (Wodarg, 26/01/10)

As with Keil's statement in the previous section, Wodarg implied here that the 2009 H1N1 strain did not actually constitute a new virus at all. Employing analogies to seasonal influenza which mirrored (though contradicted) the WHO's, the Council of Europe argued that the low mortality rate of H1N1 demonstrated that the event could not be labelled as a pandemic. Thus, 'According to the epidemiology, this swine flu was likely to be mild' (Flynn, 29/03/10). In this way it was common for critics to compare mortality rates of H1N1 and seasonal influenza, arguing that higher death rates due to seasonal influenza had been unjustifiably used as evidence by the WHO to declare a pandemic based upon spurious evidence.

The Council of Europe contested the presence of the pandemic, and the WHO's use of epidemiological statistics to justify the designation. A central claim made by Wodarg, Keil and others was that the WHO's (2009) amendments to the definition of 'pandemic' amounted to the only reason why H1N1 could constitute a pandemic. Wodarg stated: 'the current "pandemic" could only be launched by changing the definition of a pandemic and by lowering the threshold for its criteria'

(Wodarg, 26/01/10), and added that 'It is only this change that made it possible to transform a relatively mild flu into a worldwide pandemic' (Wodarg, 26/01/10). Keil stated that this occurred 'In spite of contradictory data from Mexico [the primary site of transmission] and weak and unconvincing evidence...' (Keil, 26/01/10). This aspect of the Council of Europe's account was key because it set the basis for the central claim that the WHO's alteration of the definition of a pandemic coincided with the interests of vaccine manufacturers (see below). It is important to note that this contestation represents another fundamental breakdown in the WHO's attempt to bring about scientific closure.

A second claim made by the Council of Europe's critics in relation to the definition of 'pandemic' was that, even though H1N1 could legitimately have been interpreted as a potential threat in its early stages, the WHO's announcement of a pandemic was premature. It was suggested that the WHO announced a pandemic before a true state of pandemic was in existence. Thus

> Premature announcement of a pandemic, elimination of the criteria of the level of threat of the virus by WHO and using mainly the geographic criteria without taking into consideration the number of cases actually occurring within a given region has resulted in this excessive reaction by most countries in the world...
> (Kopacz, 29/03/10)

It was argued by the Council of Europe that the WHO's actions were misplaced in declaring a pandemic. The claimed misdiagnosis by the organization regarding the state of the pandemic threat was highlighted by the Council of Europe:

> In statements made at the very beginning of 2010, WHO insisted that the world was facing a real pandemic, the future course of the pandemic was uncertain, the situation was neither overplayed nor underplayed, and the objective had always been to adopt a precautionary approach. In the same statements, WHO claimed that it was too early to say whether the pandemic was over and that another significant wave could still be expected across Europe this winter or spring.
> (Flynn, 23/03/10: 6)

This quote captures the Council of Europe's characterization of the WHO's uncertainty (as has been described in Chapter 3) regarding the future course of the pandemic. However, the Council of Europe suggested that this was not a result of the embedded risk and uncertainty of the situation but rather a result of the WHO's mismanagement. The Council of Europe questioned the judgement of the organization in declaring a pandemic and alluded to dishonest motivations behind the declaration. The pandemic potential of H1N1 was, according to the Council of Europe, not objectively evaluated by the WHO. Here it was suggested that

> When looking at the still very moderate expression of the pandemic almost one year after its outbreak (May 2010), the interpretation of scientific and empirical evidence can be seriously questioned. For some experts, it seemed obvious from a relatively early stage that the new sub-type of influenza virus was doing less harm to persons infected than other forms of the virus in previous years.
>
> (Flynn, 07/06/10: 8)

More strongly, the infectious disease specialist Rivasi said:

> I think that there are several types of responses we can have. First we have 'what is the justification of the pandemic?'. First of all, I looked at data, and in particular I looked at all the WHO alerts and reports before the pandemic was declared on the 11th of June 2009. And I think that what we find ourselves confronted with here is manipulation... It started on the 10th of April 2009 when the WHO signalled that there were flu cases in Veracruz in Mexico... Very early on Mexico, at the request of the WHO, signified that there were more flu cases...
>
> (Rivasi, 29/03/10)

Rivasi made the explicit suggestion that the WHO had engaged in manipulation by declaring the pandemic when it did – as in the quote above, where Mexico's high reporting was alleged to be a result of WHO prompting. In another example, he asserted that

> On the eve of the declaration of the pandemic, the WHO declared that the majority of cases were benign. So the cases were benign, the virus was benign, and nevertheless on the 11th of June the pandemic was declared, alert level 6. What I wondered about when looking at these facts, is the unfolding of this all. Even when we look at the WHO notifications we have the feeling that the WHO deliberately staged the events.
>
> (Rivasi, 29/03/10)

According to the Council of Europe's narrative, the H1N1 virus did not represent a pandemic threat; the pandemic declaration was unjustified. The WHO's characterization of H1N1 as a pandemic was therefore fundamentally contested in this account.

There was a fundamental institutional failure with respect to H1N1 – the WHO did not present either itself or its actions in a robust and convincing manner, leaving the 'facts' of the pandemic liable to contestation. For the Council of Europe, the WHO's actions appeared not to have been supported by scientific/'objective' evidence. The suggestedly 'unscientific' actions of the organization were presented as a key failure. For example, it was stated that

> Exactly a year ago, a very bad decision was taken by the World Health Organization that now seems unscientific and irrational. The result of that decision was that the whole world became scared that a major plague was on the way – a new pandemic that would have been as bad, according to reports, as the flu pandemic of 1918. There seems to have been no scientific basis for that decision.
> (Flynn in Council of Europe Parliamentary Assembly, 24/06/10)

Again, this suggests that the WHO defied scientific evidence in its decision-making process. However, as we have seen, the climate of scientific uncertainty under which the organization made initial decisions rendered it susceptible to such critique after the events.

The Council of Europe's depiction of the WHO's designation of H1N1 as a pandemic demonstrates that the event had been rendered liable to deconstruction. It furthermore provides evidence for the primary social constructionist claim that scientific evidence can be socially mobilized as support for primarily divergent claims. As this chapter will continue to demonstrate, both the Council of Europe and the WHO employed the same evidence basis as support for diametrically opposing viewpoints regarding H1N1.

Risk

As demonstrated earlier, the WHO emphasized the risk surrounding H1N1 and the threatening nature of the pandemic, thereby justifying the responses made. The Council of Europe presented a contradictory narrative of risk. It suggested that the WHO presented an inflated account of risk, which resulted in a disproportionate response to the threat. This followed from the Council of Europe's dispute of the concepts of 'the H1N1 virus' and 'influenza pandemic'.

In portraying the WHO's risk narrative, the Council of Europe suggested that the organization was duplicitous, or at the least inept, in its communication of risk to national governments and the general public. Thus its concern was posed:

> When looking at the still very moderate expression of the pandemic almost one year after its outbreak, the way in which scientific and empirical evidence has been interpreted can be seriously questioned. The main question is whether WHO overstated the threat posed by the virus, ignoring the practical evidence that the pandemic seemed to be of 'moderate severity' from its very start.
>
> (Flynn, 23/03/10: 3)

In this regard, it was suggested that that threat of H1N1 had been unduly exaggerated by the WHO. The WHO's reference to previous pandemics when narrating risk came under scrutiny in the Council of Europe's account. For example:

> Professor Keil ... criticised the link and references made to previously deadly influenza pandemics. In his view the comparison with the 'Spanish flu' of 1918 was generally inappropriate given the empirical figures were far from comparable. The 'Spanish flu' took place in the historical context of World War One where infections were easily transmitted by soldiers, many of whom were undernourished and without medication ... Such comparisons tended to heighten fear amongst Europeans.
>
> (Flynn, 23/03/10: 5)

However, it was acknowledged to some extent by the Council of Europe's members that the WHO was not solely responsible for this linking of H1N1 to Spanish flu and the magnification of risk. The June 2010 report asserted the weakness of such comparisons but in part absolved the WHO's responsibility for them:

> [The] WHO itself continues to assert that it has consistently evaluated the impact of the current influenza pandemic as moderate, reminding the medical community, public and media that the overwhelming majority of patients experience mild influenza-like illness and recover fully within a week, even without any form of

medical treatment. Most people, however, expected more dramatic consequences, not least because in spring 2009, the approaching swine flu was repeatedly compared to previous infectious diseases, notably the avian flu and SARS in more recent years, but also the Spanish flu of 1918.

(Flynn, 07/06/10: 12)

Here it was not directly suggested that the WHO itself fostered this image of high severity but that nonetheless the expectation of dramatic consequences had been prompted. On the whole the Council of Europe's account suggested that the WHO had constructed a discourse of high risk surrounding H1N1.

The Council of Europe was unequivocal in its assessment of the WHO's management of the risk – it argued strongly that the WHO had reacted to the threat in an inappropriate manner. Specifically, the Council of Europe emphasized the role of the precautionary principle as a determinant in the WHO's actions. The characterization of the WHO as acting primarily in this context is itself an interesting one. Though the WHO mentioned the term on a few occasions in its texts, it was by no means a reiterated concept in the organization's own account of the management of H1N1. Nonetheless, the Council of Europe continually linked the concept with the WHO's motives. Though the precautionary principle is widely considered to be a valid risk-management technique (perhaps particularly where the risk is scientifically 'uncertain', such as in the case of a pandemic) (Gollier & Treich, 2003; Liess & Hrudey, 2003), it is rendered problematic here by the Council of Europe.

The Council of Europe questioned the use of the precautionary principle in the context of H1N1. Thus it stated that

> all public health authorities concerned should critically review their way of dealing with the precautionary principle, including the communication about its use, given that the question of what society should do in the face of uncertainty is necessarily a question of public policy and not only a question of science. In future situations posing a serious risk to public health, decision-makers should bear in mind that the precautionary principle can contribute to a general feeling of anxiety and unease in the population...

(Flynn, 07/06/10: 9)

This understanding of a society-wide reaction mirrors the co-productionist claim regarding the participatory nature of contemporary science. Both scientific justification and concern for public perception are central to the Council of Europe's narrative. The concept that the application of the precautionary principle caused public anxiety was fundamental to the Council of Europe's objection to its use. Here again

> The rapporteur notes that, in some member states, the 'precautionary' approaches followed created a high degree of uncertainty and fear amongst the population, which were not necessarily justified by the evolution of the disease.
>
> (Flynn, 23/03/10: 5)

The precautionary approach, and the WHO's arguably conservative stance towards risk management in general, was thereby cast as problematic in the Council of Europe's account.

However, one interesting facet of the Council of Europe's claims was recognition that, despite the WHO's strong risk narrative, its recommendations can be (and had been) differently applied across nations. This somewhat weakens the Council of Europe's central argument that the WHO was responsible for the actions taken. It also strengthens the WHO's suggestion that responsibility was far more diffused. Thus, for example,

> on the 'precautionary principle' followed by WHO and recommended for national action, responses varied: some wished to take strong precautions, whilst others expected a lower level of outbreak of the disease, and took minimal steps. This can be seen from some of the various reactions by member states of the Council of Europe.
>
> (Flynn, 23/03/10: 2)

While most of the Council of Europe's narrative focused upon blaming the WHO, at some points it accepted that the European states were responsible for making decisions and implementing WHO recommendations.

However, generally in the Council of Europe's account the WHO had applied the precautionary principle in its management of the proposed risk in a way that led to mismanagement of H1N1. One explanation of the organization's action was that it applied the precautionary principle as a means by which to protect itself from criticism if the pandemic later proved to be severe. Thus the Council of Europe suggested strongly that 'The precautionary principle is not designed to protect decision-makers'

(Gentilini, 29/03/10). However, economic (pharmaceutical) interests were also (arguably more heavily) implicated by the Council of Europe in that

> In a situation where uncertainty is coupled with risks for human health and lives, there is also a danger that public opinion can be manipulated in favour of particular commercial interests.
> (Flynn, 07/06/10: 8)

As evident a little further below, the Council of Europe argued that the profit interests of pharmaceutical corporations were the motivation for the WHO's action. On the whole the organization's representation of risk was completely negated by the Council of Europe.

Risk and trust

The Council of Europe asserted that the WHO's mischaracterization of risk resulted in diminished trust in the management of public health. The Council of Europe's claims highlighted the centrality of trust in the institutional management of risk (see Alaszewski, 2003; Giddens, 1991; Luhmann, 2002 and others for sociological accounts). It argued that the WHO manufactured a situation which resulted in widespread panic, including, as Keil stated, 'hysterical announcements and reactions of ministries, scientific bodies and not least the media...' (Keil, 26/01/10). This panic and the associated lack of an actual threat (in terms of the Council of Europe's narrative of incidence and severity) resulted in diminished public confidence in the WHO and other public health institutions. In this way, as Wodarg claimed, 'WHO "gambled away" public confidence' (Wodarg, 26/01/10) through its handling of the incident.

As evidence that the WHO created undue public panic, the critics drew an analogy with past incidences of disease. Here, allusions were made to H5N1 (avian influenza), suggesting that this was also a case of WHO mismanagement which produced public panic and mistrust. Wodarg stated that 'there were doubts already about WHO's alarm in the avian flu in 2005/05...' (Wodarg, 26/01/10) and that 'It was then officially stated by the WHO, in panic-stricken terms, that this flu could threaten mankind and that a great number of humans could fall ill and die' (Wodarg, 26/01/10). Keil also suggested that H5N1 and other recently notable diseases such as SARS served as testament to the inappropriate way in which the WHO handled the spread of respiratory illnesses, leading to widespread concern and efforts of containment and vaccination when 'none of these pandemic predications have become true' (Keil, 26/01/10). In the Council of Europe's account, the WHO's construction of risk produced widespread panic and ultimately distrust in the organization.

Keil also made extended reference to the history of H1N1 itself, with the implication that the H1N1 subtype was an innocuous infectious agent. Here he suggested that, after the spread of H1N1 to the USA in the 1970s,

> a vaccination campaign was started in the US and about 40 million US-citizens were vaccinated because the infectious disease specialists at the CDC were convinced that H1N1 was similar to the virus that had caused that Spanish influenza... However, the H1N1 vaccination campaign was stopped abruptly when it was realized that the virus produced only a mild disease... while the vaccine produced a number of severe neurological side effects...
>
> (Keil, 26/01/10)

The critical claim was that the WHO recommended vaccination for a mild illness with no evidentiary support for its efficacy. The Council of Europe's allusions to epidemiological history also directly echoed the WHO's own references, though leading to divergent conclusions. This demonstrates both the importance of historical analogy in the social construction of disease and the potential for a fundamentally different construction using the same source 'evidence'.

The Council of Europe's documents constantly reiterated the suggestion that the WHO's actions had undermined goodwill in public health institutions. This was considered by the Council of Europe to be one of the pivotal long-term effects of the WHO's decisions with regard to H1N1. Thus it was asked:

> who will speak for the 800 million people who suffered badly as a result of this decision? And, given that we have cried wolf four times, who will suffer in the future if a very nasty disease comes along but no one believes the WHO because they no longer trust it?... We need a World Health Organization in which we can have absolute confidence...
>
> (Flynn in Council of Europe Parliamentary Assembly, 24/06/10)

The suggestion of 'crying wolf', and its detrimental effect on trust in the WHO, was prominent:

> the next time somebody cries wolf, the overwhelming majority of people will not be listening. And who do we have to thank for that? We have to thank either the inept bureaucratic dumbness of the World Health Organization or the spiteful evil manipulation of the World Health Organization by the drug companies around the world. One or other of them have to accept responsibility. If there is a pandemic in the future and people don't listen, then they [the WHO] have only themselves to blame.
> (Hancock in Council of Europe PACE Meeting, 29/03/10)

And:

> if the trust in the World Health Organization is undermined, and there have been a whole series of scares around the world, um SARS, CJD, AIDS up to a point, the millennium bug, avian flu and then swine flu. Where there have been great warnings of terrible calamities, I think, and they haven't occurred. I think the danger is, that having cried wolf so often, the public – next time there might be a real scare – there might be a virus that mutates and very few people will take notice of it. And we don't want to see the trust in the World Health Organization undermined.
> (Flynn, 29/03/10)

Employing a variety of techniques, including historical analogizing, the Council of Europe strongly argued that the WHO's actions eroded public trust. Having declared the pandemic in a time of scientific uncertainty, the organization opened itself up to the critique of 'crying wolf' when a severe threat did not eventuate.

The Council of Europe also suggested that the WHO could not be trusted to effectively assess public health priorities. Wodarg and Keil asserted that the WHO's actions resulted in the neglect of other diseases and risk factors. In this way, Keil suggested that

> Governments and public health services are only playing lip service to the prevention of these great killers [i.e. hypertension, smoking and other risk factors] and are instead wasting huge amounts of money by investing in pandemic scenarios whose evidence base is weak.
> (Keil, 26/01/10)

Wodarg too asserted that the H1N1 scare deflected efforts away from other, more important health issues. Thus the WHO's construction of H1N1 as a relevant and immediate public health threat became contested in the Council of Europe's account. In this way it was suggested that the perspective of the organization was misplaced since

> We also know that the result of the warning was that the whole priorities of health services, in any countries including my own, were distorted. Money was being spent defending against a form of flu that was very mild. Now we're simply looking after the truth, we want to find out what happened, why it happened.
>
> (Flynn, 23/03/10)

The WHO was presented as an ineffective public health institution (compare with the WHO's assertions in Chapter 7) because it was unable to manage health priorities successfully and because it did not take responsibility for its actions.

Another fundamental facet of the Council of Europe's depiction of the WHO was that the organization's actions could not be trusted because they lacked transparency. Thus

> Without transparency, suspicion remains. We are not accusing anyone of any wrongdoing, but we are entitled to know what went on. We have cried wolf four times in recent years – on sudden acute respiratory syndrome, on Creuzfeldt-Jakob disease, on avian flu and now on swine flu – and the world has been greatly alarmed, yet in all four cases, there were very few deaths around the world... There was no reason for the alarm.
>
> (Flynn in Council of Europe Parliamentary Assembly, 24/06/10)

The WHO's apparent lack of transparency was a reiterated point of the Council of Europe's narrative:

> The rapporteur is convinced that the way in which the H1N1 crisis has been handled is lacking in transparency. Certain facts have never been communicated to the European public; others have not been presented clearly enough. Even in this advanced stage of debate, and notwithstanding the lack of transparency [that] has been pointed out on various occasions, some stakeholders are still not ready to react fully to allegations made and make all possible information available.
>
> (Flynn, 2010)

Where the WHO had attempted to render itself more transparent or explain its actions, this was cast as insufficient by the Council of Europe. For example, it was argued that

> unfortunately, the testimony that we had from the World Health Organization in Strasburg [January 2010 council meeting] was not convincing. They still want to rely on secrecy and the privacy of the people involved. We don't know who took the decisions, who decided that this was going to be defined as a Phase 6 pandemic, which resulted in great alarm throughout the world.
>
> (Flynn, 23/03/10)

The WHO's announcement that it was conducting an internal review of the matter was met with similar scepticism. Almost all of the actions and responses of the WHO have come under attack by the Council of Europe in its discussions and investigations.

The Council of Europe and the WHO presented fundamentally divergent narratives of the risk posed by H1N1. Citing many of the same sources of evidence and examples as the WHO, the Council of Europe argued that the organization's mischaracterization of risk led to an erosion of trust in the institution. With regard to controlling contemporary risks, the management of public perception is crucial, due to the heavily integrated nature of the modern scientific enterprise. The Council of Europe's emphasis upon trust foreshadows the potential effect of the WHO's management of H1N1 upon its role in global public health (as discussed in depth in Chapter 8).

Pandemic phase declarations and definitions

Given the ill-defined nature of the boundary concepts 'pandemic' and 'pandemic phases', it is unsurprising that one of the strongest points of the Council of Europe's critique surrounded the WHO's pandemic phase declarations and (re)definitions. According to the Council of Europe's account, the WHO was able to portray H1N1 as a pandemic due to the fact that the organization changed its definitions of pandemic phases immediately prior to the emergence of the new H1N1 subtype. The Council of Europe argued that the premature declaration occurred because

> This declaration at a very early stage of the event... was, according to some experts, only possible because the description of pandemic alert

phases was modified by WHO in May 2009, and notably the criteria relating to the severity of the disease removed as a pre-condition for passing on to the highest alert level.

(Flynn, 07/06/10: 5)

The claim that 'scientific experts' reinforced the Council of Europe's interpretation was reiterated throughout their documents and debates. For example, it was asserted that

> A number of members of the scientific community became concerned when WHO rapidly moved towards pandemic level 6 at a time when the influenza presented relatively mild symptoms. This combined with the change in the definition of pandemic levels just before the declaration of the H1N1 pandemic heightened concerns.
>
> (Flynn, 07/06/10: 9)

The WHO's definitions of the phases was represented by the Council of Europe to have been conducted in an unscientific and unjustified manner:

> Predictions of the seriousness of the outbreak and its designation as a Phase 6 pandemic were based on a limited range of scientific opinion. Billions of dollars had been spent on the vaccine and it was necessary to clarify what had happened to avoid future repetition of the problems. The WHO had changed the criteria for a Phase 6 pandemic, basing it on this outbreak. There had been no clear answer from the WHO as to why that had happened.
>
> (Huss (representative for Luxembourg) in Council of Europe Parliamentary Assembly, 24/06/10)

Thus central to the Council of Europe's argument was the claim that the WHO (unscientifically) changed its definition of phases in order to declare an H1N1 pandemic. While the WHO referred to scientific evidence in constructing H1N1, the Council of Europe similarly enrolled scientific expertise in dismantling the organization's account.

The Council of Europe argued that H1N1 was not a 'true' pandemic and was only labelled one due to the WHO's definitional changes. The Council of Europe's position regarded the organization's statements to the contrary as further evidence of WHO manipulation and the influence of pharmaceutical interests upon the events. The Council of Europe argued that the WHO had been misinforming national governments, as demonstrated in the Council of Europe's claim that

> Although WHO continues to assert that the basic definition of a pandemic has never changed, there is watertight evidence that the former criteria... was not considered anymore in the definition used for entering pandemic level 6... the current pandemic could only have been launched by changing the definition of a pandemic and by lowering the threshold for its declaration.
>
> (Flynn, 23/03/10: 3)

And again even more emphatically in stating complicity of the WHO in constructing a 'fake' pandemic that

> It changed its criteria – when you consider it in the cold light of day and in the context of all the facts that have come out, you have to ask what the reason behind the change in the definition might have been. You cannot find anything on its website to suggest why that might have happened, who wanted it changed and on what the criteria to which it was being changed were based. There is no evidence to support that. That alone would make even the most supportive person begin to smell a rat, as they would realise that there was something seriously wrong with why such a change was being made.
>
> (Hancock (representative for the UK) in Council of Europe Parliamentary Assembly, 24/06/10)

This quote was the strongest assertion that the WHO changed the phase definitions as a deliberate ploy to manipulate the situation. Thus the 'pandemic' was described in the Council of Europe's account not as an objective entity but as a politically and institutionally constructed event, deconstructing its validity as a scientific 'fact'.

However, among the Council of Europe's statements the assessment of blame or overt manipulation lies on a spectrum. While the quote above suggests an extreme view of the WHO's liability, on the other end of the continuum it is suggested more sympathetically that

> even if WHO did not intend to modify the pandemic definition in a way that would allow for an accelerated announcement of such an event in June 2009, the changes of relevant disease descriptions and indicators at a time when a major influenza infection was already

approaching was highly inappropriate and carried out in a way which could be considered as being non-transparent.

(Flynn, 07/06/10: 10)

The fundamental claim, though, was that the WHO had changed its definition of 'pandemic' in a way that led to the mishandling of H1N1 by the organization and that H1N1 was not a 'true' pandemic.

Even more telling in terms of the WHO's lack of a stable construction was the Council of Europe's somewhat ambiguous position on the nature of pandemics and phase definitions. As noted above in the description of the nature of the virus, in parts the Council of Europe strongly asserted that the fundamental concept of 'influenza' itself is highly debateable. In the context of such a conceptualization, the definition of an 'influenza pandemic' is therefore impossible. Here it was suggested that '... if we cannot describe the ordinary (i.e. the seasonal) in any satisfactory way, we certainly cannot describe the extraordinary (i.e. pandemic)' (Jefferson, 29/03/10). Furthermore, 'This may be one of the reasons why WHO has changed the pandemic definition so many times since early May 2009' (Jefferson, 29/03/10). Within the framework of such statements it was argued by the Council of Europe that 'we can safely conclude that no one has any firm idea of how to define an influenza pandemic' (Jefferson, 29/03/10). Following from the logic of this argument, since the concept of influenza itself had been rendered contentious by the Council of Europe, the definition of a pandemic is objectively impossible. Nevertheless, in summary, it was suggested (somewhat paradoxically) that the Council of Europe

> strongly recommends that further in-depth work be done by all stakeholders concerned with a view to agreeing on a common definition and description of what an influenza pandemic is.
>
> (Flynn, 07/06/10: 10)

This again highlights the fact that definitions of pandemics (and ambiguity in the construction of these) represent one of the main points of contention within the H1N1 controversy and is symptomatic of the lack of scientific closure surrounding these phenomena. The Council of Europe's critique of the pandemic phase categories acts as further demonstration of the fragility of the construction of these definitional frames. While the phases were important in helping the WHO to define the 'thing' of pandemic, their indistinct and tenuous nature has rendered them liable to significant reconstruction and critique.

Vaccinations and other preparatory actions

The Council of Europe was interested in the issue of definitions in the context of its central concern – the WHO's management strategy, particularly the recommendation to use vaccines. H1N1 came under investigation by the Council of Europe because it was a 'pandemic whose announcement cost the world's tax payers hundreds of millions of Euros and at the same time ensured enormous additional profit for producers of vaccines for the pandemic' (Kopacz, 29/03/10). As previously argued, institutional forces within the WHO, particularly a historical dependence on vaccination, resulted in a focus on such strategies. The Council of Europe, however, suggested that the WHO's collusion with pharmaceutical corporations was the cause of this preoccupation with vaccination. Through the deconstruction of the organization's scientific fact-making, and with the added benefit of hindsight, the Council of Europe thereby questioned the motives underlying the WHO's management.

In illustrating its critique of vaccine use the Council of Europe emphasized that

> one of the central issues of the ongoing debate concerns the possibility for representatives of the pharmaceutical industry to directly influence public decisions taken with regard to the H1N1 influenza, and the question of whether some of their statements have been adopted as public health recommendations without being based in sufficient scientific evidence...
>
> (Flynn, 23/03/10: 4)

The question of the appropriate use of vaccination and suggestions of the profit-motivated influence of the pharmaceutical industry in framing the WHO's response represented the major focus of the Council of Europe's concern.

As suggested in the initial 18 December 2009 motion, the crux of Wodarg's argument revolved around the suggestion that that the H1N1 pandemic was (inaccurately) declared due to the economic interests of vaccine manufacturers. He implied that the WHO's actions had been heavily influenced by the motives of these corporations. The Council of Europe asserted that the actions of the WHO following the H5N1 (avian) pandemic, and subsequent modification of the definition of 'pandemic', were underpinned by the prospect of large financial gains by vaccine manufacturers.

Thus the Council of Europe suggested that 'As a consequence of [the] avian flu hype many contracts between national states and pharmaceutical manufacturers were signed so as to ensure the availability of relevant vaccines in case of a real future pandemic' (Wodarg, 26/01/10). These contracts were to be enacted upon the implementation of the WHO and national pandemic preparedness plans, which occurs after the WHO declares a (Phase 6) pandemic. Thus Wodarg suggested that in the case of H1N1

> The pharmaceutical companies must have been waiting for this announcement, which was made even though the flu was relatively mild. This was made possible because a new definition of pandemic levels had been adopted just beforehand.
>
> (Wodarg, 26/01/10)

It was argued that the alteration of the Pandemic Alert Phases was a result of the influence of pharmaceutical companies upon WHO actions. This argument was strongly developed in the Council of Europe's accounts. For example, it asserted that

> the credibility of an organization has been so undermined by an inability to see the wood from the trees. Or in their case being unable to differentiate between somebody paying them and worrying about where the next pandemic was coming from so to speak.
>
> (Hancock, Council of Europe PACE Meeting, 29/03/10)

And:

> We have an expression in the English language about 'who pays the piper calls the tune'. Now if there ever were to be a slogan hung over the door of the WHO, it ought to be that. With a very big question mark, the rest of you better watch out. Because it would appear that they have no scruples, do they? The evidence is apparent.
>
> (Hancock, Council of Europe PACE Meeting, 29/03/10)

These quotes demonstrate the keen interest in and the blame that the Council of Europe's members placed upon the actions of pharmaceutical corporations, and upon the WHO in yielding to their influence.

The vested interest of corporations in maximizing profit was emphasized in a number of instances. The obvious profit made by corporations was provided as evidence. Thus it was stated by the Council of Europe that

The commercial interests in the pandemic and vaccination campaigns can be illustrated by the high levels of benefit to pharmaceutical companies. According to estimations by the international investment bank JP Morgan, the sales of H1N1 vaccines in 2009 were expected to result in overall profits of between 7 and 10 billion dollars to pharmaceutical laboratories producing vaccines. According to figures presented by Sanofi-Aventis at the beginning of 2010, the group registered net profits of 7.8 billion Euros (+11%) due to a 'record year' of anti-flu vaccines sales.

(Flynn, 23/03/10: 4)

In his speech made at the March meeting, Flynn stated, slightly more charitably, that

> We did know, we do know, that there was great commercial pressure because huge (4 billion pounds) of investment had been made beforehand. So there were people who had a vested interest in making sure that huge numbers of vaccines were bought and we're not reaching any conclusion on that but I think we have to see that billions of pounds of profit were made by the pharmaceutical companies, and we're entitled to ask 'what were the interests of the people involved and who were the people involved'.

(Flynn, 23/03/10)

In this way it was almost taken for granted in the Council of Europe's account that the pharmaceutical industry strongly influenced the actions of the WHO. This contrasts with the organization's narrative which suggested that vaccine manufacturers are a responsible and necessary global partner in the management of disease threats, and that vaccines presented an efficacious and essential solution to pandemic events.

However, although the industry was often strongly portrayed by the Council of Europe as the principle antagonist in the events, its narrative also at times mirrored the WHO's characterization of the pharmaceutical industry as responsible actors. For example,

> The rapporteur also takes note of some of the reactions coming from the pharmaceutical industry. Realising that the H1N1 influenza was much milder than originally expected or feared, the pharmaceutical groups allowed many states to opt out of previous contractual arrangements and cancel orders for large quantities of non-delivered vaccines.

(Flynn, 23/03/10: 7)

Nevertheless, overall this defence was not sustained. Vaccine manufacturers were far more dominantly portrayed in terms of their financial interests:

> this test was failed also and perhaps first and foremost by companies who produce vaccines because for them, corporate profit was more important than social responsibility.
>
> (Kopacz, 29/03/10)

Furthermore, the companies' explanations of events were disbelieved because

> during the first exchange at the January hearing, the representative of the pharmaceutical industry did not provide any new evidence to dispel doubts about the possible influence that some of their members might have had on public health decisions.
>
> (Flynn, 23/03/10: 7)

On the whole the Council of Europe argued that the industry can and should act 'responsibly' but that it had been motivated heavily by the pursuit of profit. Primarily the blame was not placed on the industry as such but on the WHO because it allowed itself to be heavily influenced by the corporations whose nature it is to pursue profit.

Specifically, the Council of Europe argued that the advice to implement mass vaccination campaigns was a major error made by the WHO in the handling of H1N1 due to both the inefficacy and the cost associated with these actions. It was suggested that

> [The pandemic] declaration kicked off an immediate international agenda setting in process [including] extensive vaccination campaigns in many countries notwithstanding evidence that the influenza overall presented relatively mild clinical symptoms. In autumn 2009, several independent medical experts raised warnings regarding excessive vaccination activities for which, according to them, there was no clinical scientific evidence to justify this.
>
> (Flynn, 23/03/10: 2)

And again:

> In June 2009, the WHO declared a level 6 pandemic and vaccines were purchased in massive quantities. Without sufficient justification, 100,000 children were vaccinated. The way the pandemic has

been handled – not only by the WHO, but by the competent health authorities at European Union level – gives cause for alarm.
(Circene (representative for Latvia) in Council of Europe Parliamentary Assembly, 24/06/10)

Such statements made it clear that the Council of Europe regarded the use of vaccination as unnecessary. These contrast of course with the WHO's account which characterizes the use of vaccinations as an inevitable and essential reaction to the pandemic threat (see Chapter 5).

The suggestion that vaccination is actually effective against influenza was contested by the Council of Europe. To emphasize this, a comparison between different national vaccination strategies was made:

> Preliminary results show that there is no correlation between the amounts spent on taking precautions and the results. The country that spent the least was Poland, which rejected the idea that this disease was dangerous and which had suspicions about the safety of the vaccine... Britain spent £570 million on medicines that will never be used. The outcome, however, was that the number of deaths per million from swine flu in Britain was about twice the number in Poland.
> (Flynn in Council of Europe Parliamentary Assembly, 24/06/10)

More importantly, and central to the contestation of the conceptualization of 'influenza' and 'pandemic', the efficacy of vaccination was questioned through the use of the expert testimony of Tom Jefferson. He suggested that,

> In fact, vaccine and antivirals have a weak or non-existent evidence base against influenza. The quality of influenza vaccine studies is so bad that our systematic review of 274 vaccines studies which had [been] published between 1948 and 2007 found major discrepancies between data presented, the conclusion and the recommendation made by the authors of these studies.
> (Jefferson, 29/03/10)

The Council of Europe's account of vaccines thereby again highlighted the scientific uncertainty surrounding influenza and its management. This claimed lack of data indicating efficacy is highlighted again in the quote below:

> After reviewing more than 40 clinical trials, it is clear that the performance of the vaccines in healthy adults is nothing to get excited about. On average, perhaps 1 adult out of a 100 vaccinated

will get influenza symptoms compared to 2 out of 100 in the unvaccinated group. To put it another way we need to vaccinate 100 healthy adults to prevent one set of symptoms. However, our Cochrane review found no credible evidence that there is an effect against complications such as pneumonia or death.

(Jefferson, 29/03/10)

In addition to these allegations of a lack of efficacy, the WHO's path-dependent preoccupation with vaccination as a strategy was criticized by the Council of Europe with reference to alternatives. The Council of Europe presented the argument that broader public health measures would be more efficacious:

> Public health interventions such as hygiene measures and barriers have a much better evidence base than vaccines. They are also cheaper and socially acceptable, as well as being life savers in poor countries, yet they are almost ignored.
>
> (Jefferson, 29/03/10)

It was clear that the vaccination measures advised by the WHO presented a fundamental point of contention for the Council of Europe. The lack of a solid scientific construction of both influenza and H1N1, and vaccine efficacy, resulted in the potential for contestation of the WHO's account. The Council of Europe seized upon these fragilities in criticizing the organization's actions.

Nonetheless, simultaneously, with regard to management strategies the Council of Europe acknowledged that the advice of the WHO was to be taken as a recommendation rather than an edict. As will be developed further in Chapter 7, the WHO characterized itself as an institution which provides evidence and advice to nations but does not make decisions for governments. In the WHO's account the governments are themselves responsible. In fact, the Council of Europe's debates showed that national governments (within the European Union and elsewhere) took a variety of different actions in response to H1N1. For example, the Polish government decided not to purchase large quantities of the vaccines. These actions were explained by the Polish health minister stating that

> the conditions of purchase for vaccines proposed by producers were dubious for us, vaccines were to be purchased only by governments and not available directly to individuals, and to units of health care system, the producers of the vaccine expected that [the] Polish

government would take full responsibility for any undesirable side effects offering sale at the risk and on the responsibility of the purchaser.

(Kopacz, 29/03/10)

Thus the Polish example (and the discrepancies across European Union nations in the implementation of the WHO's advice more generally) highlighted the fact that national governments made the final decision in reacting to the WHO's declarations. Nevertheless, the WHO's role was always emphasized as the responsible and accountable agent in the Council of Europe discussions, and it was suggested that the organization 'thereby forced countries to spend billions on unnecessary supplies of medicine, as well as scaring the public all over Europe and the rest of the world' (Frahm (representative for Denmark) in Council of Europe Parliamentary Assembly, 24/06/10). Despite the potential diffusion of decision-making, the WHO was held responsible as the initiator of the situation.

The nature of the particular vaccines used against H1N1 was also a cause for criticism in Wodarg's account. As previously noted, a proportion of the vaccines (those which Wodarg critiques) used during the H1N1 pandemic had been manufactured using a new method which allows for quicker production. Wodarg suggested: 'It seems, that the indication for the new, patented vaccines primarily follow economic strategies and was not necessarily to optimise public health needs' (Wodarg, 26/01/10). Here again, he argued that economic motives were fundamental to the choice of the vaccine used. This, he asserted, was to the detriment of those who were vaccinated. Additionally, one of Wodarg's key claims was that the H1N1 vaccines were not merely unnecessary but also dangerous, arguing that the WHO acted irresponsibly in advising member states to purchase them. Due to the relatively novel method of manufacture, he suggested that the vaccines

> involved higher risks than usual vaccines against seasonal flu in [that] some adjuvants were added and injected of which we know, that they stimulate the immune system manifold, which means that they could possibly lead to autoimmune diseases (such as multiple sclerosis) and immunological complications.
>
> (Wodarg, 26/01/10)

Along with the possibility of an autoimmune response, Wodarg suggested that the vaccines might even induce cancers, asserting that

New procedures [for manufacturing the H1N1 vaccines] were allowed onto the markets to produce vaccine products including bioreactors using fast growing cancer-like cells. The possibility that their proteins could induce cancer when injected involuntarily as impurities to the patient has never been excluded from clinical testing, that needs a much longer observation period...

(Wodarg, 26/01/10)

The allusion to the possible carcinogenic nature of the vaccines is particularly interesting, and (while not strongly emphasized in Wodarg's statement at the Council of Europe hearings) it has been widely taken up by the media and other commentators (e.g. the anti-vaccination movement) (Ncayiyana, 2010; Odent, 2010; Wodarg & Villesen, 2009). Wodarg himself has been cited as having made more forceful claims of this nature to the media (see, e.g., Bancroft-Hinchey, 2010; Odent, 2010; Wodarg & Villesen, 2009). Thus, in addition to claiming that the pandemic was 'false', Wodarg suggested that the vaccines subsequently utilized in reaction to the declaration were potentially seriously harmful to those citizens who were vaccinated.

The WHO's management of H1N1 through vaccination represented an overtly political concern of the Council of Europe. The Council of Europe emphasized the role of pharmaceutical corporations and the WHO's misrepresentation of the threat. However, these claims were only made possible through the basic fallibility of the organization's construction of H1N1, the risk of pandemics, and the phase definitions. This institutional failure in establishing a solid construction led to the disintegration of the entire H1N1 actor network, rendering the management strategies open to critique.

Contested experts

As we have seen, both the Council of Europe and the WHO made reference to scientific experts in explaining H1N1. The use of scientific experts in the public management of risk has now become institutionalized. Here, 'experts' possess a key relationship to the problem at hand due to the democratized structuring of science. Experts inhabit a special status since membership of the category of 'expert' confers considerable authority and credibility (Nowotny, 2003a; Nowotny, 2003b; Nowotny et al., 2001). Furthermore, expertise is upheld not through the actions of individuals but through perceptions of the collective merit of experts as a group (Lynch, 2004; Shackley & Wynne, 1996).

Risks pose challenges to expert systems because experts must act outside their disciplinary sphere of 'expertise' in order to answer the questions that risk presents (Lynch, 2004; Shackley & Wynne, 1996; von Schomberg, 1993a). Also, importantly, the study of risk often makes use of fields in which the 'expert' may not be accomplished. For example, in the case of influenza pandemics, expert committees may consist of virologists and immunologists, as well as epidemiologists. As has been evident throughout this book, epidemiology is central to explaining H1N1. However, this field is seen as being the source of information that is not strictly objectifiable or, ultimately, authoritative. These types of 'softer' science (e.g. epidemiology, risk analysis and ecology) are less prestigious or authoritative, in many cases are less developed/'newer' disciplines, and produce results which are far more open to interpretation (Funtowicz & Ravetz, 1993). This 'objectively' indistinct nature of evidence surrounding pandemics lends to the fragility of the constructions.

The question of scientific expertise was central to both the WHO's construction and the Council of Europe's contestation of events. One of the ways in which the WHO was said to have made itself susceptible to the influence of pharmaceutical corporations was through the selection of expert committees. As illustrated in Chapter 4, the WHO phase (and pandemic) declarations and action plans were formulated in part through the use of expert committees. The Council of Europe criticized these committees, citing their lack of transparency.

The Council of Europe argued that this lack of transparency fostered situations whereby the WHO's experts might be simultaneously involved with pharmaceutical companies, leading to conflicts of interest which then resulted in the misactions. In this way it was suggested that

> Some members of these advisory bodies evidently have professional links to certain pharmaceutical groups – notably through receiving extensive research grants from big pharmaceutical groups – so that the neutrality of their advice could be contested. To date, WHO has failed to provide convincing evidence to counter these allegations and the organisation has not published the relevant declarations of interest...

And furthermore, in arguing collusion,

> It seems that the exaggeration of the pandemic was perhaps neither a mistake nor a coincidence. The pharmaceutical industries that earned

a fortune from the pandemic had their people in the WHO, which had the power to declare the pandemic and thereby oblige a number of countries to buy large supplies of products from those industries.
(Flynn, 23/03/10: 4)

As such the lack of transparency of the WHO's actions presented a reiterated point of criticism made by the Council of Europe. The institutional procedure through which scientific facts were established thereby themselves became contested.

The WHO's experts were heavily criticized in these accounts. It was suggested that 'The advisory bodies of WHO are particularly exposed to the risk of conflicts of interest regarding scientific experts' (Flynn, 23/03/10: 4). One of the main points of antagonism between the two bodies was the WHO's reluctance to release the details of the make-up of the expert committees. Thus the Council of Europe report suggested that

> The rapporteur continues to be very concerned by the lack of transparency regarding the identity of experts whose recommendations have had a major impact on public health budgets and people's health. He considers that the right of 800 million Europeans in Council of Europe member states to be fully informed should prevail over the right of a relatively small number of experts to privacy.
> (Flynn, 07/06/10: 17)

And:

> The Organization continues to hold back on releasing further information on the interests of experts, justifying this position by the need to protect experts' privacy and to prevent them from coming under extreme pressure from certain private companies or interest groups. The rapporteur is very concerned by this attitude...
> (Flynn, 07/06/10: 11)

The issue of experts had therefore become one of the most obvious points of conflict between the WHO and the Council of Europe, and it was also central to the process of institutional fact-making of both parties.

One of the reasons why the Council of Europe's members were so critical of the role of the WHO's experts may lie in their perception that such experts have undermined the politicians' own functions. This is evident in the texts analysed. For example, it was suggested that

The lack of transparency raised wider issues, such as the increasingly technical nature of issues on which politicians were required to make decisions. Experts should help decision makers, but not replace them. There was a need for the ethical questions to be considered and probably for a code of conduct.

(Huss (representative for Luxembourg) in Council of Europe Parliamentary Assembly, 24/06/10)

The insecure and contestable function of experts was thus highlighted in this debate, demonstrating the nature of 'expertise' under conditions of contemporary risk science. The public nature of expert decisions came to the fore in this case.

The Council of Europe's narrative suggested that the individuals on these committees have been deliberately misleading in their influence. For example,

Much has been said about the role of experts in advising policy makers on both seasonal and pandemic influenza. We know that some of them have been parsimonious with declaring their interests and their role as members of lobbying organizations which are financed by industry and some did not think it important to disclose pretty hefty industry funding of their institutions. We know that transparency is probably not taken very seriously by WHO.

(Jefferson, 29/03/10)

In this way the experts were presented as inherently subjective. Experts are liable to such critique due to the necessity of their inhabiting a broad multiplicity of roles within contemporary scientific knowledge production. However, for the Council of Europe, the experts are described as definitively acting in the interests of pharmaceuticals:

We have the so-called 'advisers', who offer advice to the WHO. Nearly every one of the people concerned either was or had been in the pay of one or another of the drug companies. In what other business or institution would it be possible for somebody in the pay of a body that would be the significant beneficiary of any change be able to give such unfettered advice? When the WHO received the advice, it did not even bother to challenge it.

(Hancock (representative for the UK) in Council of Europe Parliamentary Assembly, 24/06/10)

This suggested that the experts on the WHO's committees were consciously manipulating the situation in favour of pharmaceutical manufacturers, reflecting the heavily integrated nature of contemporary science which, in part, makes interaction with corporate bodies necessary.

Tellingly, the very notion of expertise was questioned by the Council of Europe, which argued that the individuals who are presented as experts are in fact manufactured entities (and thus are not 'true' experts on the subject) (for critical social scientific accounts of this nature, see, e.g., Brown, 2000; Rose & Rose, 1976). Thus

> Few realize that most experts (or KOLs – key opinion leaders – as they are known by communication agencies) do not appear like daisies in a field, they are 'made' over decades after having been recruited by specific image or communications agencies...
> (Jefferson, 29/03/10)

And:

> even experts with no ties to industry or government civil servants have career motivations, especially if they make policy and evaluate its effects.
> (Jefferson, 29/03/10)

Thus it was suggested that the use of expert opinion was fundamental to (what the Council of Europe described as) the inappropriate actions taken by the WHO. Here it was suggested

> that the result of the expert system (in which selection is on the basis of fame or sponsorship, with transparency being the exception) are plain for all to see: catastrophic predictions that have failed to materialize, poor science, a thriving pandemic industry and the reputation of public health structures in tatters.
> (Jefferson, 29/03/10)

The Council of Europe's position on experts therefore represented a clear departure from the WHO's use and characterization of such individuals, where the organization often cited expert committees in validating its claims. Nevertheless, despite this criticism of the WHO's experts, competing 'experts' were also frequently cited when providing evidence for the Council of Europe's participants' own claims (e.g. referring to the WHO acting even where expert opinion found the virus to be mild). However, though the general notion of 'expertise' was mobilized, particular experts as such are rarely mentioned, apart from the political experts who underpinned the Council of Europe's contestation.

Experts are seen as the source of objective information. However, due to the nature of risks, information surrounding a risk is necessarily tentative. This is seen in the case of H1N1, where the WHO made only heavily qualified scientific proclamations through most stages of the events. This means that in situations of risk there is likely to be disagreement among experts as to the scientific facts. In most instances of knowledge production, research occurs against a backdrop where stakeholders implicitly agree upon and 'know' (in pragmatist terms) what counts as valid knowledge (Jasanoff, 2004b). However, under circumstances of risk there tends to be greater contestations of the 'facts' of the case (Miller, 2004; Nowotny et al., 2001; Shrader-Frechette, 1993). This is because the study of uncertainty is now core to the practice of science. Thus 'Expertise is at once contested, problematical, central, and indispensable' (Nowotny et al., 2001: 215). Expert advice is sought and cited for policy-making, despite the fact that the nature of risks makes them in many ways immeasurable.

The Council of Europe's accounts strongly problematized the WHO's handling of the H1N1 threat, especially with regard to the organization's reliance upon vaccinations as a pre-emptive measure against the virus. The use of expert committees in justifying these claims was regarded by the Council of Europe as a mechanism through which the pharmaceutical corporations' influence could be fostered. This signifies one important way in which the WHO's construction of H1N1 was challenged, again indicating the instability of the construction as a whole.

Portraying the WHO's management of H1N1

Due to the fragility of the institution's constructions of H1N1, the WHO rendered itself liable to critique by the Council of Europe on basic assumptions and concepts. Overall it was argued by the Council of Europe that

> There is a great deal of evidence that the decisions were taken on an unscientific basis. We are not making accusations, but we are

entitled to transparency. There is no transparency... The only ones to benefit from the decision were the pharmaceutical companies and the vaccine manufacturers.

(Flynn in Council of Europe Parliamentary Assembly, 24/06/10)

According to the Council of Europe, the WHO's actions did not represent objective and scientifically based decisions but rather were made by a non-transparent institution which was heavily influenced by monetary interests. The weakness of the WHO's construction of H1N1, as presented in the earlier chapters of this book, meant that the Council of Europe could question its actions.

On the whole, however, the WHO was presented by the Council of Europe as an indispensable international body, but one that has made incorrect decisions in the case of H1N1. As

the World Health Organi[z]ation is the essential body in the world, it should be the health beacon for human-kind and it must assume its responsibilities and make the right choices, and there at least twice it made the wrong choices, on avian flu as well...

(Gentilini, 29/03/10)

Here the Council of Europe mirrored the WHO's accounts (Chapter 8) with regard to its role in global health:

Potential pandemics such as swine flu demonstrated the importance of having a body such as the World Health Organization, able to respond to major health threats. It was important that countries were prepared for pandemics and primed to act should there be an outbreak. It was important that countries should take preventive measures but it was wrong to force people to take such measures under the pretence of a pandemic.

(Ünal (representative for Turkey) in Council of Europe Parliamentary Assembly, 24/06/10)

In this way the work of the WHO was presented by the Council of Europe as essential. Nevertheless, the actions taken by the organization (and its lack of transparency) were heavily criticized. This placement of responsibility and blame on the WHO was the overwhelming response within the Council of Europe's discussions and documentation. There were very few examples of clear defence of the WHO's action within the

Council of Europe's documents. For instance, one French representative suggested that

> It could be true that wanting to know everything before acting meant not acting at all. It was not right to condemn the WHO, which had had to rely on expert opinions.
>
> (Rouquet (representative for France) in Council of Europe Parliamentary Assembly, 24/06/10)

However, such attempts to deflect blame from the WHO were rare in the Council of Europe's proceedings.

In some cases, wider-ranging institutional overhauls of the WHO were proposed by the Council of Europe. For example, one representative asserted that

> The WHO was an excellent organisation but it was notable that its long-term work was very good while its efforts to deal with emergencies were poor. It was a very closed organisation and there was not sufficient information about it... Transparency was the best way forward.
>
> (Huss (representative for Luxembourg) in Council of Europe Parliamentary Assembly, 24/06/10)

And another suggested that

> we believed the World Health Organization. I agree that we still have to believe them, but we must believe that the WHO will find the strength to face its own deficiencies. That is why we are sending this resolution out to the world; we do so in good faith and as an appeal. We have to face and handle all future epidemics responsibly; we must gather and act on transparent information and facts that are available to all in order to accept and estimate the degree of danger to ourselves. We should not allow ourselves to be treated as guinea pigs by anyone ever again.
>
> (Ivanji (representative for Serbia) in Council of Europe Parliamentary Assembly, 24/06/10)

Overall, the Council of Europe placed blame on the WHO for its decisions with regard to the specific problem of H1N1, and at times in its fundamental institutional aspects and alleged collusions with industry. As I have argued, all aspects of the WHO's construction and management were contested at fundamental levels. This serves to demonstrate the fragility of the WHO's construction of the H1N1 actor network. The disease was so ineffectively constructed and managed that all associated actor networks, including the WHO, were rendered liable to criticism.

Taken in the whole, the Council of Europe's descriptions of H1N1 and the influenza pandemic demonstrates that the virus failed to reach closure as a scientific fact. This is evidenced by the analysis that the Council of Europe's discussions, which show a fundamentally different description of the nature of the virus, the nature of its threat and the justified reaction to the threat. This would not have occurred if the scientific fact had been definitively established. Additionally, the manner in which the 'fact' of H1N1 was constructed (within the WHO through expert committees) came under attack by the Council of Europe, showing the fragile nature of the institutional process. Thus the Council of Europe's accounts demonstrate both that the WHO's construction of H1N1 was unstable and open to contestation, and that the WHO as an institution had become vulnerable to attack through its management of this case.

7
Globalization and Global Public Health

In providing a social scientific account of the WHO's reaction to H1N1, the organization's self-proclaimed role as a coordinator of 'global health' is key to explaining its decision-making process. The global health paradigm, which replaced earlier conceptualizations of 'international health', was fundamental to the WHO's management of H1N1. This is both because pandemics are essentially globalized diseases and because the organization strongly subscribes to the new global health perspective. It characterized H1N1 as particularly 'global' in nature. This characterization led the WHO to emphasize global cooperation and interdependence in the management of the pandemic. In respect to this global management strategy, it presented its own role as one of coordination and facilitation rather than one of action. In fact, using the lens of the global health paradigm, the WHO characterized the reaction to H1N1 as the responsibility of state governments and not its own. This distancing of responsibility was key to the WHO's narrative of H1N1. It reflects the institutional attempts to adapt to the new structuring of global health. However, the organization's positioning was somewhat ambiguous as it was perceived to be a directive body by outside actors (exemplified by the Council of Europe's narrative), where the WHO's recommendations were understood as explicit instructions. Simultaneously, the organization struggled discursively to project its role as one of coordination rather than command, despite the member states' interpretation. This struggle reflects the ambiguity of institutional roles within contemporary global public health.

The WHO's characterization of itself as being responsible for delivering information to other global health actors was also problematic. In fulfilling this role the organization rendered the processes behind the construction of H1N1 transparent. Through attempts to provide information and illuminate events, the very transparency of concepts that actually should have been black-boxed (e.g. severity) made closure surrounding the event unattainable. The obvious discursive 'constructedness' of the threat, which had come to light through the public nature of the WHO's discussion of its decision-making, rendered the WHO's account more open to deconstruction by outside actors. This attempt at transparency was a reaction to the organization's repositioning as a coordinating and information-disseminating body within the new global health paradigm. In this way the underlying globalization process, and the management of H1N1 as an outcome of the resultant restructuring of public health, was central.

This expansion of the global economy has led to greater global interdependence and had a significant impact on the effects and experience of infectious disease (Lee, 2003). The demographic changes and increased flow of people and commerce that characterize globalization have created a new vulnerability to the spread of emerging or re-emerging infectious agents, since the growth in international trade and travel facilitates the swift transmission and geographical spread of infectious disease (Lee, 2005; Woodward & Smith, 2003). For this reason, combined with a renewed political focus in the context of security, infectious diseases have recently gained greater traction as global health priorities (Ollila, 2005). The current notion of 'pandemic', mirroring ideas of global spread, reflects a particularly contemporary understanding of globalization. Both the experience and the management of infectious disease are underpinned by globalization. Furthermore, the individual and institutional perception and discourse of globalization fundamentally influence actions and reactions towards infectious disease threats (King, 2002; Petersen, 1996).

Within the social sciences, 'globalization' is a highly disputed concept. For example, many sociologists subscribe to Giddens' (1991) argument that globalization is a development which is intimately embedded within the processes of modernity, while other theorists, such as Robertson (1995), suggest that it is a trend that pre-dates modernity (Bancroft, 2001). Indeed, despite the current understanding of infectious disease as uniquely globalized, from the social history of infectious disease it is arguable that the spread of communicable agents from the 15th century onwards (through European economic and cultural expansion) was analogous to contemporary processes (Watts, 2003). Combined with the tendency for the globalizing process itself to reflect inherent contradictions, definitions of the phenomenon are often somewhat ephemeral (Bauman, 1998; Lee, 2003). However, with respect to H1N1 and its management, there are a number of aspects of globalization which are pivotal. The greater global interdependence in the management of infectious disease threats is evident. Disease events in areas which had once been spatially and temporally distant can cause disproportionate effects elsewhere in the globe (Ali & Keil, 2006). This propensity for the quick spread of disease is evident in understandings of the H1N1 pandemic. Important too is the fact that, mirrored by the pandemic itself, globalization differentially affects different nations and subpopulations but nevertheless impacts all of the global population to some extent (Giddens, 1991). Furthermore, the discourse of globalization (and perceptions about the impact of globalization) is also important because it informs management strategies.

Globalization has had a major impact on the contemporary structuring of scientific enterprise. In particular, the institutional management of globalized risk (e.g. H1N1) is explicable from a co-productionist framework. The argument is that, while the risks and the accompanying science are globalized, global politics has failed to settle into a stable institutional network. As seen in the case of H1N1, divisions of authority between global and national institutions (i.e. the WHO and national governments) is often unclearly defined (Miller, 2004), leading to confusion surrounding roles and jurisdiction (Szlezak et al., 2010). In cases such as H1N1, nation states might cede considerable power to global actors, experts and expert knowledge, but still be accountable to their citizens for the results of actions taken. This can lead to animosity between states and the managing institution (here, the WHO), as evidenced in Chapter 6 through the Council of Europe's account. In this chapter the WHO's response to this tension within global health is explored.

Regarding the question of such divisions of authority, the co-productionist investigation of issues of boundary maintenance provides some interesting insights. A pertinent argument from this perspective is the institutional ordering of these risks as 'global' in the first instance. For example, Miller (2004) argues persuasively that the 'global' nature of climate change is born from the drawing of boundaries of authority surrounding the phenomenon. Miller shows how the Intergovernmental Panel for Climate Change (IPCC) itself constructed the issue as a global one by forwarding a globalized discourse of irregular climate. The IPCC then articulated a new model of science and politics surrounding the issue – namely a global politics based upon (politically neutral) expert knowledge. This case study shows how the IPCC built institutional authority as it simultaneously constructed the globalized problem of climate change. The phenomenon of climate change exemplifies the shift of the sociocultural account of risks from local to global problems (Nowotny et al., 2001). The example of climate change indicates significant parallels to changing discourses in infectious disease, as evidenced in this chapter through the WHO's account. Miller argues that the 'globalized' nature of climate change is itself a manifestation of institutional organization. In the same way, it can be seen that historical incidences of disease (e.g. bubonic plague, smallpox or even Spanish flu) could have been considered global phenomena at the time but were rather treated as state concerns (managed by national public health regimes) due to the lack of a globalization discourse (Barry, 2004; Crosby, 1976; Zinsser, 1942). The assumption that H1N1 was a global threat, and its management as such, could be seen as a result of both a discourse of globalization and a globalized public health organization (the WHO). Furthermore, according to co-productionist theorists, the making or solving of a scientific problem lies not in a set of actions but in drawing and maintaining boundaries between multiple sources of authority. In the case of H1N1, unlike the IPCC, the institution of the WHO pre-existed the phenomenon. Nevertheless, the nature of the knowledge produced by the WHO surrounding the H1N1 risk presupposed methods through which that risk could be managed. Effectively, risks become defined in such a way as to be made manageable by (preformed) institutional structures. In this case, the pre-existing structures and aims of the WHO as a global public health institution defined the nature and management of H1N1, which was understood as a global pandemic to be managed through its recurrent strategies of mass vaccination.

The global public health paradigm

In part a result of globalization, public health has undergone significant changes in conception and organization over time. The most prominent of these is the shift from 'international health' towards 'global health'. This global public health paradigm has important consequences for the management of infectious disease threats. Brown, Cueto and Fee (2006) have demonstrated that 'global health' has thoroughly replaced 'international health' in public discourse. While the term 'global' was sometimes used before the 1990s, there are now frequent references to global health in the discourse, with allusions to 'international health' declining (Brown et al., 2006). This shift is not only semantic but it also reflects wider structural changes. The term 'global health' emerged as a consequence of the impact of the broader historical, economic and political processes that are embedded within globalization. 'International health', which referred to the control of epidemics across the boundaries of a nation state, was the predominant concept during the 19th and 20th century. In contrast, 'global health' implies the needs of a global population which supersede the interests of individual nation states (Brown et al., 2006; Yach & Bettcher, 1998b).

In a discursive sense, the global public health paradigm is important because of the way in which this institutionalized discourse implicitly contains value-laden suggestions of the (proper) organization of health systems. The concept refers to a consciousness that the world is a single networked space, which in turn implies political assumptions about how public health should be ordered (Keane, 1998). The predominance of 'global health' suggests that public health issues should be understood and managed on a global scale. This understanding changes the role of the nation state, and of transnational actors, as the domestic and global spheres of policy and action become entangled (Walt, 1988; Yach & Bettcher, 1998). These changes also reflect tangible alterations in the structure of public health. Due to the impacts of globalized interdependence, the number and scale of health concerns (particularly in the context of infectious disease) is growing (Taylor, 2005). Infectious agents can indeed move more swiftly across the globe, rendering national boundaries meaningless in the management of pandemic disease (Brown et al., 2006; Buse & Walt, 2000; Janes & Corbett, 2009; Taylor, 2005). This has created an emphasis on global health governance. Critically, this has changed the nature and role of the WHO.

There has been a move from 'international governance' towards 'global governance' in the management of disease threats (Brown et al., 2006; Fidler, 2004). International governance, the past structuring of public health, reflected governance structures focused on the sovereignty of the nation state, and included the association of intergovernmental agencies, such as the WHO (as it was then conceptualized) (Brown et al., 2006; Taylor, 2005). Contemporarily, global governance refers to the repositioning of state actors and intergovernmental (now global) organizations, and the inclusion of a range of non-state actors such as NGOs and multinational corporations (Brown et al., 2006; Buse & Walt, 2000; Maguire & Hardy, 2006; Taylor, 2005). Health policy is now formed at the global level through networks of private-public partnerships. Some commentators have particularly emphasized the changing role of the private sector, where private actors (including and especially pharmaceutical corporations) have gained increasing power over the governance of public health (Buse & Walt, 2000; Taylor, 2005).

Importantly, public health management has become increasingly fragmented and verticalized (Ollila, 2005), with emphasis being placed upon selected interventions (particularly with respect to infectious disease) through a growing number of public-private partnerships.

It is clear that the WHO subscribed to the understanding of public health as 'global'. The notion of 'global health' came to the forefront at many stages of the WHO's discussions. This can be evidenced most directly in that the H1N1 threat and associated reactions were frequently referred to by the WHO using the specific term 'global' (this is evident throughout this chapter). The shift towards a global worldview is apparent in the following examples:

> In the face of this, WHO strongly emphasizes that continued global cooperation is really the essential basis for fighting this pandemic. And not just this pandemic but also future health challenges.
> (Fukuda, 03/12/09)

This quote shows the basic understanding of public health responses as inherently 'global' in nature. The extract below further demonstrates the effect of this characterization. Here the characterization of the problem as global is manifested in the management through global partnerships.

> We actively embrace the idea, that working with a broad coalition of partners, in this instance really a global coalition of partners, is essential for handling these kinds of threats. Now this approach is definitely necessary for the current pandemic, but I think it's also clear that it's going to be necessary for the future global health threats as you can appreciate I think, that we have been a very highly connected and fast-moving, globalized world right now, and WHO considers that working in isolation is not really an option.
> (Fukuda, 03/12/09)

The WHO understood the phenomenon of H1N1 through the lens of global health and managed it accordingly. This perspective was integral to key decisions that it made.

The global threat

The global health paradigm implicitly rests upon the perception of the effects of globalization. The H1N1 threat, like many widespread infectious disease events, had clearly been described as a globalized disease. The WHO's narrative reflected an understanding of H1N1 as global in nature. For example, one of the aspects of the virus that was most heavily emphasized was its ability to cross boundaries and affect diverse populations. This formed the mechanism through which the risk was characterized as global in nature. Thus

> Influenza pandemics, whether moderate or severe, are remarkable events because of the almost universal susceptibility of the world's population to infection. We're all in this together, and we will all get through this, together.
>
> (Chan, 11/06/09b)

Furthermore, in addition to H1N1 being characterized as a global threat, the WHO also suggests that the pandemic itself was the result of globalization. Here it was suggested that globalization increased the potential impact of H1N1 in that

> The world today is more vulnerable to the adverse effects of an influenza pandemic than it was in 1968, when the last pandemic of the previous century began.
>
> (Chan, 11/06/09)

This is because

> The speed and volume of international travel have increased to an astonishing degree... The radically increased interdependence of countries amplifies the potential for economic disruption [caused by pandemic disease].
>
> (Chan, 11/06/09)

In this way the notion of globalization was prevalent and was referred to in the texts in order to convey both risk and an understanding of the need for cooperation.

Characterizations of the globalized nature of the threat also occurred through the WHO's linking of H1N1 with other global disasters, notably the 2008/2009 global financial crisis. In analogy to that crisis, H1N1 was described as 'another global contagion' (Chan, 11/06/09) and it was suggested that

> these crises come at a time of radically increased interdependence among nations, their financial markets, economies, and trade systems. All of these crises are global, and will hit developing countries and vulnerable populations the hardest. All threaten to leave this world even more dangerously out of balance.
>
> (Chan, 11/06/09)

Globalization and Global Public Health 179

In this way the idea of global interconnectedness was used to reinforce the notion that H1N1 could hold significant implications worldwide – and this global nature in itself characterized the virus as a risk.

Moreover, infectious disease threats in general were represented as highly globalized. The fear of an influenza pandemic was therefore (at least partially) represented as a consequence of a fear of globalized threats. Past global infectious disease threats were invoked by the WHO in relation to H1N1, and with regard to consequences of such global threats:

> What the SARS and avian influenza epidemics both showed is that when this new kind of threat can appear, they can threaten large numbers of countries in many different ways, not just the disease, but the fear these diseases can have effects on economies, on societies, and...the world is really interconnected at many different levels...And so these new emerging infectious disease threats are truly international and global [in] scope.
>
> (Fukuda, 28/04/09)

This statement shows the strong link between H1N1 and globalization. The idea of global spread is key to the risk surrounding the pandemic, an event which mirrors the fear and distrust surrounding the process of globalization itself (Bauman, 1999; Beck, 1992).

In this way the H1N1 threat was clearly characterized within discourses of globalization. This understanding of H1N1 as a globalized threat was fundamental to the characterization of management and the roles of various key actors within the global public health structure. The WHO's reaction was underpinned by these understandings of globalized risk.

The role of the WHO

At key points in its history, the WHO has led, reflected or adjusted to changes in the wider structuring of public health (Brown et al., 2006). The recently changing context of public health necessarily resulted in shifting governance structures, including shifts in the structures and practices of the WHO. In fact, the rise of the global public health paradigm was deeply influential in the institutional arrangement of the WHO. Principally, the WHO's structures changed as a reaction to the appearance of new players in the global health arena (Kickbusch & de Leeuw, 1999; Maguire & Hardy, 2006; Szlezak et al.,

2010; Taylor, 2005). Prior to the late 1990s, the WHO had been recognized unquestioningly as the leader of international health. However, by 1998, it was seen as an organization in crisis (Brown et al., 2006). The dominance of global health had resulted in the diminishing of the WHO's status. New actors, such as private corporations and global NGOs, had risen up and implicitly challenged the WHO's authority over the management of public health policy and its actions (Brown et al., 2006; Szlezak et al., 2010). As a reaction to this, the WHO began to change its role to suit the new global health environment. Instead of presenting itself as a key decision-making body, it began to reconstruct itself into the role of coordinator and strategic planner. This is clearly evident (below) in the WHO's narratives of its own role and actions.

The tension between globalization as a lived reality and its governance is clear in this case study. The erosion of the jurisdiction of the nation state, and the rise of health problems which transgress national boundaries, left authority over public health increasingly ambiguous (Szlezak et al., 2010; Taylor, 2005). This tension has given rise to new institutional forms, including shifts in the WHO's own structures. Some commentators have suggested that we are currently experiencing a flux in institutional arrangements, as the management system transitions into a more authentic 'global health' situation (Szlezak et al., 2010). The H1N1 example appears to indicate that this is indeed the case since part of the WHO's difficulty with regard to the pandemic was the ambiguity of its new role. Currently, public transnational organizations such as the WHO serve as mechanisms for the facilitation of multilateral cooperation and action (Taylor, 2005). This allows for the WHO to negotiate arrangements between diverse stakeholders and to facilitate global action. In this way it has shifted from an authoritative force to acting increasingly as a coordinating body. Its ability to fulfil a directive leadership role had been based upon the political support of its member states (and especially those that supply the bulk of the funding) in the international health paradigm (Taylor, 2005). In the context of global health, the input and effect of increasing non-state actors has diluted this initial mandate.

The WHO now perceives itself as primarily concerned with the coordination and facilitation of dialogue among various global public policy networks, which include not only state actors but also corporations, NGOs and other elements of civil society. Thus, although some authors have suggested that increasing interdependence strengthens the role of organizations such as the WHO (particularly due to their perceived neutrality) (Taylor, 2005; Walt, 1988), there has been an overall weakening of authority which has relegated the organization to a 'facilitator' rather than a leadership position. The rise of authority in global public-private partnerships (and the effective exclusion of the WHO as a determining force in some of these) has distanced the influence of the organization (Buse & Walt, 2000; Kickbusch & de Leeuw, 1999; Szlezak et al., 2010; Taylor, 2005). In this way, diverse actors within global health, including in this case pharmaceutical corporations, are treated as 'partners' in accordance with the new paradigm (Buse & Walt, 2000; Ollila, 2005), and this is evident in the WHO's narratives of such actors (refer to the depiction of vaccine manufacturers in Chapter 5).

Within this new structure, the WHO has put itself forward as responsible for managing and disseminating public health information, and organizing global partners during times of crisis. These assumed roles within global health have been specified and strengthened in the revised 2005 International Health Regulations (IHRs). These specify the legal obligations of both the WHO and its member states in relation to the management of public health. The 2005 version reflects a shift towards global health through the recognition of the erosion of state sovereignty in this area, and an increase of the jurisdiction of the WHO in its 'coordinator' capacity (Baker & Fidler, 2006; Mack, 2006/2007). This came in the form of an emphasis on global surveillance, where states are now under an obligation to notify the organization of all events which may constitute a global health problem (Baker & Fidler, 2006). The WHO is then responsible for organizing the reaction to this reporting. In this way it has positioned itself as the primary coordinating global health body. This presumed role within the new global health system is evident throughout its references to its own actions.

The new structuring of the WHO as a result of the rise of global health is evident in the management of H1N1. Fundamentally, the WHO depicted itself as contributing to global health primarily through coordinating diverse public health organizations and governments. Its self-adopted role was thus to coordinate global efforts against disease – coordination and assistance were emphasized as opposed to delivering recommendations or engaging in direct action. This distinction was illustrated throughout the texts in suggestions such as those below:

> this is a time in which we can work with countries to be as prepared as possible. That is the bottom line. Our bottom line is that there

are things that countries can do, that we can help them with, to get them prepared for this kind of potential increase in people getting sick. And this is why we are so serious about this event.

(Fukuda, 07/05/09)

This quote suggests that the role of the WHO was to assist countries in preparation, not to make decisions in and of themselves. This is evidenced again below:

> This is one of the core areas where WHO typically spends a lot of its efforts, trying to identify from country to country [their capacities and resources], what are the needs there, and then to bring together the international community. So this may mean working with donors, it means working with technical partners. It means working with all of those different entities out there that can provide help – UN organization sisters and so on.
>
> (Ben Embarek, 04/05/09)

> and say what is most important, the most important things are that, countries are as prepared as possible. This is a single most important action and this is a single biggest help that WHO can provide to countries.
>
> (Fukuda, 02/05/09)

These quotes emphasize the importance placed by the WHO upon coordination and information-gathering rather than in action. The adherence to principles of global public health is thus clear. They also show that the countries, not the WHO, are liable for the implementation of protective measures.

The role of coordination is also apparent in the WHO's narratives of pharmaceutical corporations and other private-sector actors. Instead of managing these actors, or providing direction to them, the organization presents itself as simply bringing the stakeholders together:

> A third parallel process related to vaccines is very close contact between WHO and other public health agencies and with the private sector, with the manufacturers out there. One of the things we are simply trying to do is that in this kind of extraordinary situation, make sure that the public sector and private sector are very well coordinated. So that they understand what are the priorities for the public health side and we understand what are the priorities and realities for

the private sector, for the manufacturers. This is where there really has been an extensive amount of discussion and collaborate work between vaccine manufacturers and public health.

(Fukuda, 22/05/09)

This narrative suggests that coordinating (not instructing) these actors is the primary goal, emphasizing the 'partnership' nature of global public health.

However, although the purely coordinating role was held as ideal, it was not consistent in the discussion of all contexts. At a few particular points the WHO presented itself instead as a vital actor and decision-maker. For instance, in narrating the general mobilization in reaction to H1N1, it portrayed itself as the responsible actor in the face of global emergencies. For example, in one reference to the morale of staff during the development of the H1N1 threat it was stated:

Now, having said all that we are tired, the odd loud word is said, but what we have is had lots of practice unfortunately, with SARS, with tsunamis, with major responses to epidemics. We vaccinate millions of people every year in response to meningitis epidemics, we can move millions of vaccines and we can mount mass campaigns to vaccinate people, we can contain outbreaks of Ebola in the rain forest.... In SARS we got very tired and many of us appeared to have reached burn-out, this time we intend to be able to maintain this pace for as long as is necessary to provide our public service to our Member States and to communities.

(Ryan, 02/05/09)

Here the WHO is forcefully portrayed as an important and active agent in managing health crises. This is presented again here, where it was asserted that:

this is our business really, and WHO mobilizes to handle sudden emergencies. We do this very often, whether this is Ebola (haemorrhagic fever) in Africa or the Tsunami spread over a very wide area. Some countries fortunately can deal with a crisis once in a century. As Mike pointed out we [the WHO] deal with 250 events a year. And that isn't just reporting an event, that is responding to an event.

(Ryan, 02/05/09)

In total:

> In a sense really being prepared for public health issues is a never ending job. Because the diseases change, the scope of the problem changes, the world changes and public health has to keep up with it. The bottom line message is that the kinds of dangers we face are changing in the modern world. Of course public health has to change to keep up with it. It is a kind of dog race.
> (Fukuda, 07/05/09)

The organization's perception of itself as actively working in a struggle against infectious disease (more synonymous with its previous, more central role within structures of international health) was presented here. However, in general the 'active' potential of the WHO was rarely emphasized.

References to itself as an active decision-making agent were rare in the organization's texts. Instead, in general, the WHO minimized any suggestion of responsibility for the events. In this regard, the case of vaccines is again pertinent. It can be argued that the WHO is the primary agency for making decisions regarding which vaccines are manufactured and which viral strains are focused upon. This is because it monitors and releases data about which strains are prevalent and are considered potential threats. However, this responsibility for vaccine manufacture was not acknowledged by the WHO in the case of H1N1. Instead, the lack of authority over the use and implementation of vaccines was constantly emphasized (even before their use had been widely criticized). The WHO positioned itself as a source of information rather than advice. This is a pivotal distinction. It demonstrates a key aspect of the global health paradigm – responsibility (like risk) is spread across a multitude of actors and stakeholders, including the WHO and national governments, but also the media and industry. This dissemination and diminishing of ultimate responsibility was emphasized in the WHO's texts.

Instead of making decisions, the WHO considered itself as primarily providing information. This position was highlighted by statements such as the following:

> I think that the job of public health is really to alert the public when there are significant dangers to which they may be exposed and then also to identify the options and the things that people can do to protect themselves against that danger. For example, with the pandemic situation, getting useful information, accurate information out to the

populations is one of the basic jobs to public health and this is both true for national groups as well as for WHO.

(Fukuda, 17/12/09)

This quote provides a clear indication of the role that the WHO has adopted. As with the management of most risks (as suggested by Beck, 1992, 1999), information is socially perceived to be crucial to harm minimization, and the WHO positioned itself as a critical organization in the management of globalized risks by suggesting that it provides access to vital information.

The global health paradigm was emphasized through the organization's allusions to the collaborative nature of risk management. In addition to coordinating other public health bodies, the WHO's actions were depicted as a result of these multiple perspectives. Thus, for example, although the director-general appeared to take responsibility when she suggested that 'The decision to declare an influenza pandemic will fall on my shoulders [and] I can assure you, I will take this decision with utmost care and responsibility' (Chan, 08/05/09), there is also a distinct sense in which the position of the WHO was dependent upon the actions of member states and other stakeholders, such as pharmaceutical corporations. In this way, Chan simultaneously asserted that she 'will follow your [national health official's] instructions carefully ... in discharging my duties and responsibilities to Member States.' Furthermore, the input of multiple partners was emphasized. For example, in announcing the decision to call a pandemic, Chan suggested that the organization had 'conferred with leading influenza experts, virologists, and public health officials' (Chan, 11/06/09). This impression of the WHO's actions as being dependent upon and a result of the input of multiple individuals, governments and organizations was clearly distinct from the narratives of critics and commentators more generally, who tended to portray the WHO as solely responsible for making the decision to call a pandemic and dictating reaction. It also lends to the primacy of the globalized public health paradigm, which constructs reactions to global health threats as interdependent upon the actions of multiple stakeholders.

As a whole, the importance of global public health was reinforced throughout the texts. The notion that public health is a neglected area was also often highlighted. Thus it was suggested that

> Time and again, health is a peripheral issue when the policies that shape the world are set. When health policies clash with prospects

of economic gain, economic interests trump health concerns time and again. Time and again, health bears the brunt of short-sighted narrowly focused policies made in other sectors.

(Chan, 11/06/09)

And furthermore:

All [of the present global crises] will show the consequences of decades of failure to invest in health systems, decades off failure to consider the importance of equity, and decades of blind faith that mere economic growth is the be-all, end-all, cure-for-all. It is not.

(Chan, 11/06/09)

In this way the WHO perceived its handling of the H1N1 pandemic as critical both to producing increased attention to public health and to managing perceptions of its own institutional relevance. Chan suggested that 'How we manage this situation can be an investment case for public health' (Chan, 11/06/09). The H1N1 pandemic was therefore perceived as pivotal to the wider perception of global public health and the role of the WHO.

The WHO narrated its role, then, as being a champion in the cause of global public health and a coordinating body within these structures. Critically, this served as a measure to diffuse responsibility across multiple actors, as the WHO was depicted as coordinating actors rather than an organization which presented edicts that determined action. This role signifies the shift in global health, where globalized cooperation is understood as the mechanism through which global risks should be managed.

Globalization and cooperation

The shift towards global health carried important consequences for the structuring of public health actions. The increased emphasis upon coordination and cooperation was central to the WHO's discourse of health management. The United Nations system as a whole began to collaborate increasingly with private interests towards the end of the 20th century for a variety of practical and political reasons (Ollila, 2005). Combined with the discourse of global health, this meant that the nature of public health shifted radically, with the rise of global public-private partnerships (GPPPs). These denoted a shift away from nation-based policy-making towards the increasing collaboration of private partners (Buse & Walt, 2000; Janes & Corbett, 2009; Ollila, 2005). The traditional actors within public health – the WHO and nation states – were thereby being joined (and challenged) by a growing number of elements within civil society (including NGOs, corporations and religious groups) (Reinicke, 1999; Szlezak et al., 2010). In fulfilling its role of coordinating body, the WHO must emphasize the continued potential for cooperation between these diverse actors. It is clear that GPPPs reflect an increasing interdependence between a variety of state and non-state actors. Furthermore, there are changing relationships between the actors, such that the formal and informal norms and expectations have become vague (Szlezak et al., 2010). This has created challenges for the WHO in terms of coordination. One way in which the WHO had attempted to negotiate this was through its discursive practice, constructing the problem in such as way was to render it manageable. In the case of H1N1, the organization repeatedly insisted on the importance of partnerships and cooperation in the conduct of public health policy.

The WHO's texts strongly suggested that the reaction to the threat must be a global one. Corresponding to the discourse of global public health, it was asserted that the threat of H1N1 affected all nations and, furthermore, that the reaction to the threat should be multi-institutional and cooperative. Thus, in keeping with the proposed universal nature of the threat, the concept of 'global solidarity' was key to the WHO's depiction of necessary action against H1N1. It was emphasized that 'An influenza pandemic is a global event that calls for global solidarity' (Chan, 04/05/09) and that

> An influenza pandemic is an extreme expression of the need for solidarity before a shared threat... As I said, an influenza pandemic is an extreme expression of the need for global solidarity. We are all in this together. And we will all get through this, together.
>
> (Chan, 11/06/09)

The suggestion that 'we are all in this together' was characteristic of the WHO's depiction of the necessary global reaction to H1N1. In this way, the notion of worldwide vulnerability and the importance of global cooperation was often emphasized through the WHO's account.

The specific term 'global solidarity' was heavily utilized, particularly throughout the director-general's speeches (indicating the organization's most important and public announcements). It was suggested that 'All countries profit from this expression of solidarity' (Chan, 18/05/09), and the idea of working in cooperation was emphasized throughout. The following quotes illustrate the strong discursive use of the concept of solidarity:

> Above all, this is an opportunity for global solidarity as we look for responses and solutions that benefit all countries, all of humanity. After all, it is really all of humanity that is under threat during a pandemic.
>
> (Chan, 29/04/09)

And:

> Constant, random mutation is the survival mechanism of the microbial world. Like all influenza viruses, H1N1 has the advantage of surprise on its side... We have another advantage on our side... collaboration and solidarity.
>
> (Chan, 17/08/09)

As these extracts suggest, though the H1N1 virus was depicted as capable of significant disruption and harm, the notions of a common humanity and 'working together' against the virus was invoked as an important protective mechanism. In accordance with its coordinating role within global public health, the WHO emphasized cooperation between actors as a means by which to combat the pandemic.

Mirroring ideas about globalization and global health, it was asserted by the WHO that global cooperation was a key to managing infectious disease threats. Examples of cooperation were celebrated:

> I would like to say that we have seen, if you compare this to previous events, we have seen a remarkable amount of openness and transparency and cooperation between countries.
>
> (Ryan, 02/05/09)

In this way the WHO narrative stressed the importance of global collaboration, mirroring the emphasis of global public health. Thus

> Calling a pandemic is also a signal to the international community. This is a time where the world's countries, rich or poor, big or small, must come together in the name of global solidarity to make sure that no countries because of poor resources, no countries' people should be left behind without help.
>
> (Chan, 11/06/09)

This emphasis helped to sustain the WHO's role as coordinator of global public health efforts and o provide continued meaning to its work, despite its loss of authority and its previous standing within international health.

Both the WHO's characterizations of H1N1 and reactions to the disease emphasized the concept of global public health. The WHO represented itself as a coordinating body which provided a source of global information. With regard to its narrative and practice of global public health, the practical implications of the blurring of the roles of various stakeholders were evident. A good illustration of these implications was the organization's reaction to pharmaceutical manufacturers. The 'global' and cooperative nature of vaccine manufacture was emphasized in the WHO's accounts. For example, it was suggested that

> Development of these actions each involved working with a range of global partners, and this is a general principle that we follow at WHO: to be as inclusive as possible. One of the specific actions taken by WHO was to focus on vaccines.
>
> (Fukuda, 03/12/09)
>
> making and distributing and administering the pandemic flu vaccine was going to be very complex, difficult and time-consuming task. So from the outset it was clear that we would have to be working with multiple partners, both in the public and private sectors... Given these considerations, we did move quickly to mobilize these global partners.
>
> (Fukuda, 03/12/09)

As these quotes show, allusion to global cooperation was one way in which the WHO upheld its vaccination strategy. The role of vaccine manufacturers in this collaborative context was emphasized thus:

> this is one of the key ways in which the public sector and the private sector work together on global health problems. This kind of collaboration is really essential for dealing with a disease like influenza because the information comes from countries through their monitoring and assessment activities and then the vaccines come from the private sector because that is where the manufacturing capabilities are. What we try to do is facilitate and make this process as effective as possible.
>
> (Fukuda, 11/02/10)

...maintaining and engaging the private manufacturing sector has been a very critical step, again, because this group has the unique and essential role in the vaccine manufacturing process.... In the first place it's the private sector which makes vaccines... Also, this group that has really a unique expertise and knowledge of vaccines because of their manufacturing of the vaccines, it's essential for public health really to act on this kind of knowledge and know-how....
(Fukuda, 03/12/09)

These quotes illustrate a variety of ways in which the WHO narrated the use of pharmaceuticals. These included emphasis upon the 'expertise' and 'knowledge' of corporations in this area and the designation of the private sector as 'partners' in an 'inclusive' manner. This worked to characterize the WHO's role as one of facilitation, distancing perceptions of the organization as the sole responsible actor. These narratives all fit in with the global public health paradigm, which focuses not just on the WHO and nation states but on other global actors, including corporations.

The global public health paradigm and the association of the pandemic with the process of globalization therefore had a strong effect upon the way in which descriptions and reactions to H1N1 were mobilized. In accordance with the new global health, the WHO positioned itself as a coordinating agent. Multiple institutions were therefore conceptualized as partners in the efforts against H1N1. More generally in reference to globalization, a global and coordinated (rather than national) effort was characterized as pivotal.

The relationship of the WHO with national governments

The emphasis on 'solidarity' and treating all actors as 'partners' had important flow-on consequences. One key effect of globalization, and the shift towards global public health, is the changing role of the state. Generally, a significant trend of the globalization process is the increasing influence of supranational organizations (Bauman, 1998). Public health in the West had historically been associated with the needs of national security and commerce, where health policy was based upon the assumption that national governments could control what occurs within their own borders (Bashford, 2002; Bashford & Strange, 2003; Bauman, 1998; King, 2002). However, the globalized nature of infectious disease spread diminishes state capacity to internally manage public health (Kickbusch & de Leeuw, 1999). The present 'global' nature of public health therefore represents a subversion of state jurisdiction. Though the degree to which the state is threatened remains a subject of intense debate within sociology (Lee, 2003), the reality of contemporary infectious disease does suggest significant erosions in territorial power by restricting the policy-making capacity of governments (Fidler, 2001; Szlezak et al., 2010; Taylor, 2005).

Global health governance is primarily concerned with facilitating multilateral cooperation among nation states and non-state actors (Taylor, 2005). However, importantly, while health appears to be necessarily an area for global action, due in large part to its interrelation with security, it remains a policy and management area which nation states protectively guard (Kickbusch & de Leeuw, 1999). There is therefore a tension between the global nature of infectious disease spread and the desire of national governments to control health. In terms of the WHO's characterization of H1N1 and suggested preparatory actions, this retention of state sovereignty over health has led to uneven implementation of global policy. While globalization and global health tends to weaken the role of the nation state, national governments are ultimately responsible for the implementation of global policy into domestic law and action. The tension between state and the WHO's accounts was therefore clear, and evident in the texts in several instances. Foremost was the critique made by the Council of Europe. However, the tension was also evident within the WHO's own accounts of the H1N1 pandemic.

Globalized public health has led to significant implications surrounding the relationship of the WHO with its member states. As demonstrated throughout this chapter, the WHO positioned itself as an institution that was concerned with collecting and disseminating information rather than providing decisions which determined actions. This was evidenced most starkly in the relationship of the organization with national governments. In several areas the WHO suggested that it acted as a source of information and not action. For example, in explaining of the utility of the Pandemic Alert Phases, it was suggested that

> This entire planning process was really initiated to help countries develop their preparations as much as possible so that in the advent of a pandemic they would be better off than they would be without the process. So the pandemic Phases are really a planning tool for countries and a way to alert them that there is a situation that they need to be aware of and as a tool to make sure that they understand as

we go into different Phases, there are different actions which should be considered by them and some of them which should be taken.

(Fukuda, 22/05/09)

Here the phases were characterized as a planning tool which provided information to member states, as opposed to concrete statements of action. Again, in the context of characterizing the H1N1 threat, it was stated that

> Basically we have this list of indicators and we use them to assess, first the disease itself, and help countries to assess their own vulnerability. Rather than a guidance, I would say, it is more a concept paper plus some operational tools to make best use of the information we have and to better support countries in planning.
>
> (Fukuda, 13/05/09)

Again, the WHO provides information but the national governments act. On the whole, the WHO did not consider itself to be in a position to offer recommendations to individual countries, but rather suggested that it acted as a source of global information. Governments themselves could choose (how) to act on this information. On one level, this was justified by the WHO's argument that it focused upon the global condition of the threat, and therefore that national governments must assess individual national responses. Thus, for example, with regard to severity (made when the concept was still utilized), it was asserted that

> Severity can be taken in two dimensions: at the global level, that is what WHO is doing, we are reviewing the situation in different countries within the World Health Organization, and we give a global assessment. But we would encourage each country to look at their own situation to make a national assessment on severity; and in continental countries – big countries – they may even consider looking at what would be the severity at sub-national level.
>
> (Chan, 11/06/09)

And again:

> There are different local risks and there are different global risks, so each individual event must be assessed in its own merits and we will be assisting countries with the advice they need to make those decisions.
>
> (Ryan, 02/05/09)

Thus, by positioning itself as a global body, the WHO's responsibility for local actions was absolved. As the quote above clearly suggests, for the organization it was the countries themselves which make decisions in terms of monitoring, risk assessment and ultimate action.

The WHO thus emphasized the independence of national governments in forming reactions to the threat. In some instances throughout the texts, this was stated explicitly. For example, it was suggested that

> governments do not necessarily wait for WHO to make recommendations before they do anything and in fact many governments are very proactively working on the situation now... On the other hand, I know that many governments are also looking at what their plans are if the situation escalates and what possible actions they may take. So I think the governments are being very active right now and they are certainly not being passive. Nonetheless, I think they are looking to WHO for guidance...
>
> (Fukuda, 26/04/09)

As this quote suggested, the organization depicted itself as providing evidence, and to some extent 'guidance', whereas the governments themselves were portrayed as being responsible for decision-making. The emphasis, then, was upon the autonomy of individual nations to make choices for their citizens. Thus, for example, in response to a question regarding the vaccination of entire populations and whether the WHO would 'think it realistic and do you suggest to [other] governments that they should do the same...' (Keiny, 06/08/09), the representative answered in terms of the individual country's autonomy:

> Some countries have decided to vaccinate their whole population – there is no indication that this would be unsafe so it is again a strategy of a country to protect its population against influenza pandemic. Not all countries which could have access to enough vaccine have chosen to do this, again it is... the country's choice... the choice of a population to be vaccinated is a national prerogative and each country will have to take this decision in view of their own epidemiological and national characteristics.
>
> (Fukuda, 24/09/09)

There was tension, then, between a global health paradigm (coordinated by the WHO) and an international health paradigm (managed by individual nation states). The international health paradigm rests upon the protective actions of nation states, whereas the global health paradigm rests upon globalized cooperative action. While the WHO narratives emphasize the important of globalized action, in practice the onus of decision-making is still constructed as a national-level event, and the organization actively distanced itself from decision-making. This shows that the roles of the diverse actors within global health remain in flux.

Furthermore, the distancing of the WHO from the actions of nation states resulted in an important unintended corollary – that is, that the organization did not perceive itself to be responsible for the actions of nations and, furthermore, suggested that it was not in a position to scrutinize the actions of state governments. For example, it was asserted by the WHO that

> Earlier on in this series of press conferences, I said that one of the things I didn't want to do is comment on a particular action being taken by any one country... There are very difficult issues for national authorities to weigh. I think it is a little bit hard from outside, simply to say: these are good or bad actions. They are very difficult actions... because of on the other hand it turns out many people are very severely ill and they were not jumping on it early, they will also be criticized. I will just stop here and say that these are very difficult issues that the governments wrestle with and of course they try to make the best decisions that they can, given the information they have.
>
> (Fukuda, 07/05/09)

It is clear from this quote that the WHO did not wish to portray itself as being accountable for the results of management decisions. It depicted itself as responsible only for providing information and facilitating dialogue. This statement shows that it attempted to distance itself from national action, even though criticism from the member states and the Council of Europe cited the WHO as the responsible agent.

This detached response to the actions of governments can also be illustrated in specific examples. One was the actions of the Norwegian government, which early on had made the anti-viral oseltamivir available over the counter. The dominant infectious disease perspective on anti-viral use suggests that overusage can directly result in anti-viral resistance (Hayden, 2006; Patel & Gorman, 2009). However, the WHO did not criticize this action, even though it could have had widespread (global) consequences which could be reasonably argued to be part of the organization's jurisdiction. Here it was suggested that

> we have been in close contact with the Norwegian authorities both to find out about the situation in the country and to discuss whether there is anything that WHO can offer them. One of the interesting things which the Norwegians are doing is to make antiviral drugs more easily available for a limited period of time. The reasons they are doing this is that the stress on the primary health care system is quite high...
> (Fukuda, 05/11/09)

In this statement, a conciliatory tone is evident and it was clear that the representative had sought to evade any evaluation of the Norwegian government's actions. In other instances, with regard to allegations of favouritism, misinformation or misbehaviour on the part of national governments, the representatives again remained distant in their observations. For example, in response to a question about inaccurate reporting:

> We think that the national health authorities do report accurately to the WHO. As I am sure you know, to confirm a death has been caused by H1N1 needs some confirmation and therefore we may receive it a little bit later but we are confident that the reporting that we get is what is really happening.
> (Kieny, 19/11/09)

Here the detached tone which the WHO adopted in relation to the actions of national governments was again evident. Likewise, with regard to the question of whether there might have been bias in vaccine distribution in some countries, it was answered: 'We hope not! The governments are usually very responsible for that' (Fukuda, 25/09/09).

In this way the WHO's narrative placed the burden of responsibility primarily upon individual nations, and portrayed itself as a source of (objective/scientific) information and a facilitator/mediator of the different stakeholders present in the global public health arena. This served to distance responsibility from the WHO. However, blame for mismanagement was placed on the organization regardless, as seen in the narrative of the Council of Europe. The combination of these depictions shows that the role of actors within global public health was yet to be consolidated, a factor which contributed to the instability of the H1N1 construct.

Developing countries

The boundaries of authority surrounding the management of H1N1 were indistinct. Although the interdependent and cooperative nature of global health can serve to eradicate boundaries, one of the inherent contradictions of globalization is that it blurs but can also reinforce borders (Bashford & Strange, 2003; Woodward & Smith, 2003). The way in which pandemics spread across previously spatially defined borders is evidence of the blurring effect. However, the definition of space and the maintenance of boundaries are often fundamental to long-held social mores, and boundaries can be strongly protected (Bashford & Strange, 2003). This is evidenced not only in the division of authority between the governments and the WHO but also in the reaction of developed world governments to the developing world. In the case of H1N1, the WHO found that it needed to defend the rights and actions of developing nations. This was particularly evident with respect to discourses of isolation and quarantine. The wider problem of pandemic management reflects contradictions between the ideal of global cooperation and the tendency to reinforce boundaries between the developed and developing worlds.

It is important to note here that public health priorities often reflect the concerns of the wealthy (Ollila, 2005). In this case the advent of a global pandemic would rate as a priority for the West whereas the developing world faces more pressing immediate concerns (despite the fact that a pandemic would affect the developing world, with its lack of health infrastructure, disproportionately). Simultaneously, infectious disease problems are often perceived as originating from the developing world. Fundamental to these perceptions are what King (2002) refers to as the 'emerging disease worldview'. This has arisen in the West and narrates the link between the developed and the developing world through the experience of infectious disease. Here the subjective perception of globalized interdependence is linked with moral narratives locating disease in the Third World to construct a discourse which suggests that the West is increasingly susceptible to infectious disease threats which originated in developing countries (King, 2002). The tendency to see H1N1 as located and arising from the developing world is evident in the WHO's texts, where the organization counsels developed nations against taking drastic actions against (the citizens of) developing countries.

Though the blaming of the developing world is in itself an important aspect of the sociology of infectious disease (Abeysinghe & White, 2011; Bashford, 2002; Foege, 1991; Nelkin & Gilman, 1991), the exploration of this area is not within the scope of this book. What is important in the context of the present discussion is the way in which the WHO managed this blaming. Historically, the WHO has portrayed itself as a champion of the interests of developing nations, and this was also evident in the case of H1N1. It should be noted here that while globalized diseases have the power to affect all nations, some are unequally impacted. Thus the director-general stated that 'It is my duty to help ensure that people are not left unaided simply because of the place where they were born' (Chan, 04/05/09) and she strongly urged wealthy nations to 'look closely at anything and everything we can do, collectively, to protect developing countries from, once again, bearing the brunt of contagion' (Chan, 11/06/09).

Throughout the texts, the representatives emphasized 'the absolute need to extend preparedness and mitigation measures to the developing world' (Chan, 11/06/09) in part because of the unequal impact that pandemic influenza might have under the health conditions found in such regions. Thus it was stated that

> Although the pandemic appears to have moderate severity in comparatively well-off countries, it is prudent to anticipate a bleaker picture as the virus spreads to areas with limited resources, poor health care, and a high prevalence of underlying health conditions.
> (Chan, 17/06/09)

As 'we do not know how this virus will behave under conditions typically found in the developing world' (Chan, 17/06/09), the WHO emphasized its potential effect upon developing nations. Again, the notion of variable global severity is highlighted here. Furthermore, the image of the WHO as the protector of developing nation's interests coincided with its wider public goals.

One of the WHO's fundamental goals was to attempt to ensure equitable health outcomes. Where developing countries were referred to, the representatives emphasized the responsibility of the WHO in reducing the vulnerability of these populations. Thus

> One of the important tasks at this point is to anticipate that the needs of countries if we go into that situation. In particular, what we are really focusing on, or beginning to focus on, is the anticipated

needs of developing countries if the pandemic should develop and if these countries get impacted. We know from history, we know from the analysis of past pandemics, and we also know from many infectious disease and health problems that the poorer and the developing countries are the ones who really get hit the hardest.

(Fukuda, 28/04/09)

Furthermore,

certainly some developing countries are more vulnerable in a sense that they have a high proportion that is malnourished and that are probably, is more, let us say, fragile for this particular disease.

(Fukuda, 13/05/09)

In this way, though it is arguable that the 2009 H1N1 strain had not placed a huge health burden on affluent nations (and could be reasonably referred to as mild), the global perspective of the WHO might justify its concern over the disease to some extent. The organization argued that it was difficult to predict how the spread of the virus would impact upon developing nations. For example, it asserted that

perhaps of greatest concern, we do not know how this virus will behave under conditions typically found in the developing world. To date, the vast majority of cases have been detected and investigated in comparatively well-off countries.

(Chan, 11/06/09)

And:

Although the pandemic appears to have moderate severity in comparatively well-off countries, it is prudent to anticipate a bleaker picture as the virus spreads to areas with limited resources, poor health care, and a high prevalence of underlying medical problems.

(Chan, 11/06/09)

In this way, although critics in affluent nations may have disparaged the actions of the WHO, from the perspective of global health, and particularly the health of developing countries, it was arguable that the H1N1 strain may have caused a dramatic impact in poorly resourced areas.

In contrast with other sources of public discourse, which can tend to situate developing nations as scapegoats for the spread of disease, the WHO's perspective described such nations in the context of profound inequalities. Its advocacy of the interests of developing countries was evident in the discussion of vaccines and anti-virals. Here it was noted that

> in total [the] WHO's global stockpile [is] up to 10 million treatment courses... But we don't think that this is enough to meet the needs of the countries. So we have been working with partners and also with other countries who have enough supplies to meet the global need.
>
> (Shindo, 12/11/09)

Thus the 'WHO is really trying to ensure that all countries, including developing countries, will have access to vaccines' (Kieny, 06/08/09). Overall, the question of equity was thus central to the WHO's reaction. With regard to this it was acknowledged that vaccines would not be fairly distributed:

> Who will get the vaccine? Well, of course the first countries to receive the vaccines will be two categories of countries. First are the rich countries, with a high income. These are the ones which have already at the beginning even before the pandemic started, purchase agreements with manufacturers... The other type of country to be served very early with the vaccines is the countries that they do not have to be rich, but to have domestic production [e.g. China, which at this point in time had already started mass vaccination campaigns].
>
> (Fukuda, 24/09/09)

Thus 'a final point that I want to make about vaccines is that we are in a situation in which some countries have vaccine available and other countries do not' (Fukuda, 05/11/09). As such, the WHO asserted that it coordinated with manufacturers and more affluent nations to ensure a more just distribution on a global scale. The representatives suggested that the 'WHO is negotiating with the manufacturers to have access to vaccines for developing countries and this is through donations or purchase at a reduced price...' (Kieny, 06/08/09) and that it was

> in line with this and are discussing with manufacturers about having access to their production capacity... on behalf of developing countries we are really striving to make sure that the quantity of the vaccine that WHO will be able to access directly, not talking about

what these countries negotiate themselves, will cover at least these populations.

(Kieny, 06/08/09)

The question of advocacy on the part of developing nations corresponded to the positioning of the WHO within the wider global health arena. However, this image of the developing world as being burdened by contagion and vulnerable to the actions of the developed world contrasts with dominant portrayals of the developing world as the source of contagion, as presented by developed countries.

'International hostilities'

The actions of developing nations were in fact questioned by media queries to the WHO's representatives. For instance, during the early part of the H1N1 threat (though decreasingly as the event went on) there was considerable fear of travellers from developing countries. It was suggested by governments of developing nations in several cases that they had been subject to discrimination through other nations' disease control measures. For example, as the first cases of H1N1 arose in Mexico, that country quickly became a target of sanctions. The WHO, however, failed to react. For example, one reporter (Eva Ussi, Grupa Radio Centro, Mexico) asserted that

> the influenza virus has already caused international hostilities particularly against Mexico, who have seen how Argentina, Cuba, Ecuador and China have cancelled flights to and from this country. China even went further and kept Mexican businessmen and Mexican tourists, around 70 people, secluded in a hotel. They were not infected, nevertheless they could only return to Mexico on a special flight. Mexicans feel hurt because they were unilaterally stigmatized for being Mexicans. This treatment was not given to the United States or Canada [which at this point also had presented cases]. This attitude actually contradicts the recommendations of the WHO, doesn't it?
>
> (Fukuda, 06/05/09)

To which it was answered that

> countries can take additional measures, other than those recommended by WHO that they feel might be necessary to respond to a public health risk. However, countries adopting measures that are significantly different and/or interfere with international traffic must

provide WHO the public health rationale and relevant scientific information for those measures. We have begun the process of getting more information from a number of countries... We do remind you that the IHR does require that Member Countries treat travellers with respect for their human rights, dignity and fundamental freedoms.

(Fukuda, 06/05/09)

In a second instance, the reporter (Frank Jordans, Associated Press) stated that 'we have heard a lot about discrimination against people from certain countries because there are outbreaks there' (Briand, 08/05/09), to which it was replied:

according to the International Health Regulations a country, if it wants to take health measures above and beyond what is recommended by WHO can do so, but it must justify those in public health terms. Often times, WHO will write to a country asking for justification for these measures. We have done that in quite a few instances already, I don't exactly know how many, and we have received responses...

(Briand, 08/05/09)

These quotes suggest, as illustrated elsewhere, that the representatives took great care not to engage in discussion regarding the actions of specific nations and only responded with reiterations of the general actions taken by the WHO with regard to such cases. The described role of the organization in upholding the interests of the developing world therefore clashes with the general stance of disinterestedness, as this example illustrates.

The institutional positioning of the WHO was an important factor in the reaction to the perceived threat. The H1N1 pandemic was both a globalized disease and a product of the perception of globalization. The risk was perceived through the understanding of H1N1 as a consequence of the globalized world. Furthermore, both its characterization and its management were dependent upon the WHO as an institution being heavily influenced by the change to a paradigm of global public health. Acting within this paradigm, the WHO demonstrated a strong subscription to its self-construction as a global coordinator. Combined with the emphasis of the global health paradigm upon cooperation between multiple global actors, it had framed itself as the institution responsible for facilitating global cooperation to combat globalized disease threats. The organization depicted itself as providing information to global actors but not giving direction or creating expectation. It thereby both distanced itself from responsibility and upheld the primacy of global public health. While the WHO continues to depict itself as a champion of the developing world, this non-directive positioning results in an inability to criticize the actions of member states.

The reaction to H1N1 as a whole was therefore informed through an understanding of the impact of the global health paradigm upon the contemporary management of health threats. Due to the fact that the move towards global health has left the roles of key actors in flux, the rise of the global health paradigm contributed to the contestation of H1N1. Here, in addition to the weak construction of the facts of H1N1 (as demonstrated throughout this book), the ambiguity of the roles and responsibilities of key actors within the framework of global health led to a crisis of management. As Chapter 6 demonstrated, nation states understood the WHO as the responsible agent, while (as shown throughout the present chapter) the organization perceived ultimate responsibility as resting upon national governments. This confusion in roles contributed to the unstable management of H1N1.

Another potentially important sociological point with regard to global health is the transparency of the process. The successful construction of a scientific fact requires the erasure not only of ambiguity surrounding the phenomenon but also of the producer. The producers of the fact need to be erased from the representation of the fact so that the process of construction is hidden (Derksen, 2000; Fleck, 1979; Latour & Woolgar, 1979; Lewin, 1994). Along with its other weaknesses, this may be one of the reasons why the WHO's construction of both the pandemic phases and the H1N1 virus came under contestation – the organization made its actions of construction transparent to the member states and the media. The documents analysed in this book are testament to the fact that the WHO provided a detailed discourse of construction, instead of masking its actions and presenting artefacts such as the pandemic phases and H1N1 as incontestable scientific realities. This is partly due to the effect of risk upon the production of scientific research, as articulated by the co-productionist theories, and partly due to the coordinating role of the WHO within global public health. Contemporary science is more likely to be open to public engagement due to the shifting structures of research (Funtowicz & Ravetz, 1993; Jasanoff, 2004b). The WHO was a victim of this open discourse, and this resulted in the deconstruction of its classificatory scheme by outside actors. Its role as being responsible for disseminating information, as prescribed by global public health, may in fact have rendered the construction of H1N1 more susceptible to critique.

The WHO's depiction of global public health, and its place within it, was fundamental to its reaction to H1N1, and in part its lack of success in relation to public perception of the pandemic. The WHO understood itself as a coordinator but, as evidenced in Chapter 6, it was widely perceived as giving directives for national governments. Furthermore, the WHO understood its role as one of facilitating dialogue and disseminating information, but in reality it was held to be much more responsible for the results of the H1N1 campaign. The ambiguity of the WHO's role within the emerging structures of global public health thereby underpinned the fragility of the construction of H1N1 as a whole.

8
Conclusions

The case study of H1N1 shows that the initial definition of a disease event can have wide-ranging impacts. This act of definition was underpinned by complex social mechanisms which collectively formed an attempt to constitute a pandemic event as a scientific fact. Though most such facts appear as objective realities, since scientific closure has been achieved, in some cases the attempted construction of a fact is fragile and unstable, leading to their contestation. In such cases, the socially constructed nature of scientific facts becomes more apparent; the assumptions behind the phenomenon become unravelled through the act of contestation. Prior to the event of H1N1, it was generally assumed by key scientific and institutional stakeholders (including the WHO) that a pandemic event could easily be identified. However, the case of H1N1 problematized this black-boxed understanding of a pandemic as an objective and readily distinguishable scientific reality. This was evidenced by the contestation of outside actors, including the prominent voice of the Council of Europe.

Due to the instability of the WHO's account of H1N1, the event was vulnerable to contestation. The concept of the H1N1 pandemic became susceptible to critique as a result of the institutional and wider social context in which the construction was produced by the WHO. The instability of the 'fact' of the H1N1 pandemic was a result of the embedded scientific uncertainty surrounding H1N1, the lack of pre-existing clarity surrounding the concept of 'pandemic', the institutional processes of the WHO, and the positioning of the WHO within global public health. This underpins the broader argument that contemporary global risks are defined by uncertain science, which risk-managing institutions must negotiate.

For scientific closure to occur, where the phenomenon is unproblematically accepted by all actors within the relevant actor network, a disease event needs to reach stability and incontestability as a scientific fact. Through examining this case study we saw that the WHO's attempt to produce scientific closure surrounding the H1N1 pandemic had limited success. The stability of the concept of the H1N1 pandemic was compromised in a number of ways. These included the lack of early consensus on the name (important in defining and distinguishing a scientific artefact), the failure to produce a coherent disease narrative (essential in producing a socially significant discourse of disease threat), and ineffectual comparisons with seasonal influenza and historical pandemic events (key to producing analogies that tap into collective memories of infectious disease). This shows that, while the WHO subscribed to important and well-worn cultural mechanisms for producing a disease narrative (as described in the sociology of health and illness literature) in the construction of H1N1, this was conducted in an ineffectual manner. This lack of an initial robust definition of the H1N1 virus was pivotal to the eventual instability and contestation of the WHO's account.

This ineffectual construction led to the problematizing of the concept of 'pandemic' itself. In describing the H1N1 virus as a pandemic threat, the WHO translated the virus into the actor network of the concept 'pandemic'. The idea of 'pandemic', previous to H1N1, had been a black-boxed concept whereby all actors believed that they understood what was meant by the term. However, H1N1 did not correspond to these prior assumptions of 'pandemic'. The WHO's attempts at translating the H1N1 virus as a 'pandemic' were therefore problematic. The instability of the construction of H1N1 led to the general problematization of the concept of 'pandemic'. As such, the translation of an unstable scientific fact into an actor network can have the effect of problematizing the other elements of that actor network. In this way, through the attempted translation of H1N1, the basic notion of a pandemic became subject to contestation.

This was particularly problematic for the WHO, in that once the pandemic declaration was made it was necessary for the organization to continue to uphold the designation of H1N1 as a pandemic. Another important consequence of this labelling was the construction of the H1N1 pandemic as a significant risk. In order to construct H1N1 as a legitimate problem, the WHO needed to mobilize a risk discourse surrounding the event. It failed to create an effective risk discourse surrounding H1N1. Further, the H1N1 virus was characterized by a lack of scientific data, making the (then) future course of the pandemic difficult to predict. The WHO reacted to this shifting foundation in scientific evidence by co-opting the idea of scientific uncertainty into its risk narrative. While these explanations may have attempted to reflect the inconclusiveness of scientific evidence surrounding H1N1, they did not aid in the construction of a persuasive risk narrative.

Effective risk narratives are consistent and convey a clear sense of threat. The WHO's risk discourse surrounding H1N1 failed to fulfil these functions. Scientific institutions (in this case the WHO) need to subscribe to one of the many possible scientific accounts surrounding a phenomenon in order to makes sense of it, despite the existing scientific uncertainty. However, the WHO's account was ambiguous about the risk surrounding the event, as evidenced by its narratives of statistics and uncertainty. Further, where it did choose between competing scientific explanations, the account that it favoured was incompatible with the pre-existing and prevalent understandings of 'pandemic'. A pandemic event is understood as a risk because of its severity. In defining severity (which was, like 'pandemic', another previously unproblematic concept), the WHO chose to emphasize the criterion of geographical spread. However, this understanding was challenged due to the fact that H1N1 failed to result in high morbidity and mortality, despite the geographical spread. This, too, rendered the WHO's construction of H1N1 open to contestation, as the risk discourse surrounding the event was both inconsistent and ineffectual.

The failure to construct an effective risk discourse was in part a result of the previous institutional understanding and experience of pandemics, as evident in the WHO's Pandemic Alert Phases. The act of categorization was crucial to the events surrounding the H1N1 pandemic. H1N1 was ultimately contested by outside actors, including the Council of Europe, on the grounds that it did not constitute a genuine pandemic threat. The WHO's placement of H1N1 into the category 'pandemic' was central to the critiques. Institutionally, the WHO, as the defining actor, had attempted to formalize this tacit understanding of 'pandemic' in the pandemic phases. The phases describe categories of an influenza threat and define a 'pandemic' event. However, as the concept of pandemic was so implicit, the phases were ill defined and described in ways that did not readily correspond to wider social understandings of the nature of pandemics (particularly with regard to severity). This contributed to the fragility of the construction of H1N1 as a pandemic and the contestation of the WHO's depiction of events.

However, though the construction of H1N1 as a pandemic threat was fragile, it was nonetheless sustained by the WHO. Due to this, following the construction of the threat, attempts at resolution were necessary. In addition to identifying and providing information surrounding pandemic events, the WHO is responsible for managing global action aimed at controlling the impact of pandemics. It emphasized mass vaccination as the most effective strategy against H1N1. This emphasis upon vaccination was a result of the path-dependent institutional reaction of the organization, where prior experience with infectious disease (including notable victories using mass vaccination strategies) resulted in the favouring of this reaction in the case of H1N1. Other potential actions were disregarded or underemphasized. However, this came under contestation when H1N1 did not manifest into a significant disease threat. This again formed an important element of the critique of the WHO's management of H1N1.

The fragilities of the WHO's construction of H1N1 are most easily discernable through the analysis of the critique of its account. Contestation often has the effect of revealing the mechanisms through which the contested scientific fact was constructed. This was clearly the case with respect to H1N1, where previously black-boxed concepts such as 'pandemic' and 'severity' were rendered problematic through their association with the failed construction. The Council of Europe fundamentally contested the definition of 'pandemic', the legitimacy of the WHO's risk narrative surrounding H1N1, the organization's definition of pandemic phase categories, and the management strategy emphasizing vaccine use. In so doing, by illustrating the ineffectual construction of the global event of the H1N1 pandemic, the legitimacy and institutional processes of the WHO were rendered susceptible to critique by the Council of Europe. This again demonstrated the forceful and widespread impact that a contested construction can have upon its actor network.

Both the Council of Europe's reaction and the WHO's handling of H1N1 were results of the wider context of global public health in which the organizations were acting. The recent shift from international public health to global public health changed the roles of key actors in health management, particularly the WHO. The rise of global public health was characterized by the addition of multiple actors and stakeholders in the health arena. It was also characterized by a shift towards understanding these diverse actors as 'partners' in improving public health. Within this shifting climate, the role of the WHO changed from a directive body in advising public health actions to a coordinating body that was more responsible for distributing information surrounding health risks. From the WHO's perspective, it was concerned with providing information and promoting dialogue surrounding H1N1, not with prescribing the reactions of national governments in managing the pandemic. However, there was an evident confusion in roles here, as the Council of Europe account clearly focused upon the WHO as a directive body, the dictates of which, it was suggested, produced unnecessary action (and cost) for national governments. The indistinct nature of roles within contemporary global public health resulted in confusion on the part of the WHO with respect to the type of information that it should be producing, and consternation on the part of the governments represented by the Council of Europe which still perceived the WHO as a managing and directive institution.

The 'facts' of a disease can be produced in such a way as to render the construction unstable, leading not only to the contestation of the fact itself but also the instability of associated components of the actor network. In this case the concepts of 'pandemic' and 'severity', and the legitimacy of the institutional processes of the WHO, were all problematized or critiqued as a result of their association with the fragile construction of H1N1. The social structures surrounding the construction of the H1N1 pandemic, including the nature of global public health, institutionalized reactions to infectious disease, the need to co-opt black-boxed concepts, and the lack of consistent scientific empirical evidence, combined with the deficit of manifestation of severe clinical disease, all resulted in a lack of scientific closure and the construction of a socially fragile scientific fact. In recent years there has been increasing emphasis upon the potential threat of global pandemic disease. This perception has resulted in a range of institutional and societal reactions, which provides an important point of interest for the burgeoning sociology of infectious disease as well as the analysis of other global threats.

This study does not either validate or criticize the WHO's handling of H1N1. Rather, it argues that social representations and institutional structures are central to the way in which infectious disease threats are perceived and managed (and are themselves a product of social forces). Despite the influence of constructionist social sciences, contemporary infectious diseases are generally understood as 'real' objective threats. This is mirrored by the social science surrounding respiratory diseases, which tend to be demographic, focus upon lay risk perceptions, or examine political and governance mechanisms. In contrast, this case study shows that social representations and constructions are pivotal in framing the reaction to a disease. These are evident within social action when a threat becomes contested, such as occurred in the case of H1N1. However, they are equally vital, though perhaps less apparent, in cases where consensus surrounding the representation of disease has been achieved.

Moving away from infectious diseases, it is clear that contemporary 'wicked problems', such as climate change, population growth, and food and water scarcity, present important similarities to pandemics. All are global risks of a potentially catastrophic magnitude, and all are similarly often framed by scientific uncertainty and contestation. Just like the WHO with respect to H1N1, global institutions and national governments are placed in a position where they are forced to act, even while scientific evidence is scarce or conflicting. Understanding the role of scientific uncertainty therefore presents an important contribution to analysing these problems through a social scientific lens. Representation and social construction are fundamental to the way in which these risks are perceived and managed. As Latour (2004) put it, such risks are not 'matters of fact' to be taken for granted, even though they are global 'matters of concern' which must be confronted. Understanding the fact-making surrounding these concerns is therefore pivotal to reflecting upon the management of global threats.

Appendices

1 Timeline of Events

Outbreak of H1N1 Detected

18 March 2009: Surveillance conducted by the government of Mexico begins picking up the first cases of influenza-like illness

24 April 2009: The WHO releases its first statements on the virus. This is the first report of influenza-like illness outbreaks caused by A/H1N1 in the USA and Mexico
- The majority of cases in otherwise healthy young adults

Phase 3 Pandemic Alert

25 April 2009: First meeting of the WHO's Emergency Committee is held:
- Agreed that the situation constitutes a public health emergency of international concern
- All countries to intensify surveillance
- More information is necessary before deciding on the appropriateness of raising the alert from the current Phase 3

Phase 4 Pandemic Alert

27 April 2009: 73 laboratory-confirmed cases of H1N1 worldwide:
- Mexico (26 cases including 7 deaths)
- USA (40 cases)
- Canada (6), Spain (1)

Second meeting of the Emergency Committee is held:
- Based upon epidemiological data suggesting human-to-human transmission and the ability of the virus to cause community-level outbreaks, the influenza pandemic alert is raised to Phase 4
- Containment is not considered feasible
- Not recommended to close borders or restrict international travel
- Production of seasonal influenza flu vaccine to continue
  – WHO to facilitate the process needed to develop an A(H1N1) vaccine

Phase 5 Pandemic Alert

29 April 2009: 148 laboratory-confirmed cases and 8 deaths of H1N1 over 9 countries worldwide, including:
- Mexico (26 cases including 7 deaths)
- USA (90 cases including 1 death)
- Austria (1), Canada (13), Germany (3), Israel (2), New Zealand (3), Spain (4) and the UK (5)

Appendices 211

	Influenza alert level raised to Phase 5 – All countries to activate their pandemic preparedness plans
30 April 2009:	257 laboratory-confirmed cases and 8 deaths of H1N1 over 9 countries worldwide, including: – Mexico (97 cases including 7 deaths) – US (109 cases including 1 death) – Austria (1), Canada (19), Germany (3), Israel (2), Netherlands (1), New Zealand (3), Spain (13), Switzerland (1) and the UK (8)
8 May 2009:	Special meeting of the ASEAN+3 health ministers (in Bangkok) to address the H1N1 pandemic
18–22 May 2009:	The 62nd World Health Assembly convenes in Geneva. H1N1 and the Pandemic Alert Phases phases are an important discussion point
29 May 2009	53 countries, over all continents, have reported 15,510 cases of H1N1, including 99 confirmed deaths

Full-Scale (Phase 6) Pandemic

10 June 2009:	74 countries have reported 27,737 cases, including 141 confirmed deaths
11 June 2009:	WHO Director-General Margaret Chan declares that H1N1 constitutes a Phase 6 pandemic
1 July 2009:	77,201 cases worldwide, 332 confirmed deaths
31 Jul 2009:	162,380 cases, 1154 confirmed deaths
20 September 2009:	Over 300,000 cases, including 3,917 confirmed deaths
23 September 2009:	Emergency Committee holds its fifth meeting – no amendments to recommendations
25 October 2009:	Over 440,000 cases, including 5,700 confirmed deaths
26 November 2009:	Emergency Committee holds its sixth meeting – Amendment to recommendation on travel: because pandemic infections are widespread, there is no longer any scientific reason to delay travel to reduce the spread of infection.
4 December 2009:	8,768 confirmed deaths
18 December 2009:	Motion recommended to the Council of Europe by Wodarg and associates: 'Faked Pandemics: A Threat to Public Health?'
26 January 2010:	Public Hearing of Council of Europe: 'The Handing of the H1N1 Pandemic: More Transparency Needed?'
5 February 2010:	15,174 confirmed deaths
29 March 2010:	Council of Europe PACE Meeting on the WHO's handling of the H1N1 pandemic

2 May 2010:	18,001 confirmed deaths
7 June 2010:	Council of Europe's report on the handing of H1N1, prepared by Paul Flynn (UK) is released. It is highly critical of the WHO's actions
24 June 2010:	The 26th sitting of the Council of Europe's Parliamentary Assembly debates the WHO's handling of H1N1. The assembly passes Wodarg's 18 December motion
27 June 2010:	18,239 confirmed deaths
1 August 2010:	18,449 confirmed deaths

Post-Pandemic Period

10 August 2010:	WHO Director-General Margaret Chan declares the end to the Pandemic (beginning of a Post-Pandemic Period)

[End of period under analysis for this thesis]

2 Key Actors

World Health Organization

Margaret CHAN:	WHO Director-General
Keiji FUKUDA:	Special Advisor to the WHO Director-General on Pandemic Influenza; Assistant-Director General. Fukuda acts as the WHO's spokesperson for the majority of media releases and public statements
Marie-Paule KIENY:	Director of the Initiative for Vaccine Research
Nikki SHINDO:	Medical Officer in the WHO's Global Influenza Programme
Harvey FINEBERG:	Chair of the International Health Regulations Committee
Peter BEN EMBAREK:	WHO Food Safety Scientist
Michael RYAN:	WHO Director of Global Alert and Response

Council of Europe

Paul FLYNN:	UK Socialist – elected as rapporteur for the inquiry into H1N1
Wolfgang WODARG:	Epidemiologist/physician and then-current (2009 – though was not re-elected) parliamentarian of the Council of Europe (for Germany)
Ulrich KEIL:	Epidemiologist, Director of the WHO's Collaborating Centre for Epidemiology and Prevention of Cardiovascular and Other Chronic Diseases at the University of Munster

Tom JEFFERSON: Scientist from the Cochrane Institute
Marc GENTILINI: Infectious disease expert
Ewa KOPACZ: Minister of Health for Poland
Michèle RIVASI: Member of Parliament, Group of Greens/European Free Alliance

Pharmaceutical Corporations

Luc HESSEL: Representative of the European Vaccine Manufacturers present at the Council of Europe\s enquiries

Institutions and Committees

GOARN: (WHO) Global Outbreak Alert and Response Network, the technical collaboration of international resources in outbreak surveillance and management

IHR: International Health Regulations (2005), which provide a framework for coordinating the management of international public health emergencies

PACE: Parliamentary Assembly of the Council of Europe

3 Summary of the WHO's Pandemic Alert Phases[1]

PHASE	WHO's DESCRIPTION	WHO's RECOMMENDED ACTIONS
1	No animal influenza virus which causes infections in humans	None (preparing response plans)
2	Known infection of humans from animal influenza viruses	None (preparing response plans)
3	Animal or human/animal virus causes sporadic cases or small clusters	None (preparing response plans)
	No human-to-human transmission	
4	Human-to-human transmission occurs OR	Rapid containment measures
	Human-animal virus demonstrates ability to sustain community-level outbreaks	Readiness for pandemic response
5	Virus causes sustained community outbreaks in at least two countries in one WHO region	Each country implements its national pandemic response plans
6	Pandemic in progress	Each country implements its national pandemic response plans
	Virus now causes community outbreaks over multiple WHO regions	
Post-Pandemic Period	Levels of influenza have returned to those corresponding to normal seasonal influenza activity	Evaluate the response and revise plans

Notes

3 Risk and Scientific Uncertainty

1. The lack of an articulated definition of 'severity' even within the scientific literature further points to the black-boxed nature of the term. Within such literature, for example, different types of severity exist – that is, case-load severity, or a relative severity index in respect of a particular disease. These terms all imply a basic (black-boxed) understanding of 'severity', but the term itself is ill defined, or defined only by a tautology (see Porta, 2008).

Appendices

1. The table has been derived from the 'Pandemic Influenza Guidance and Response: A WHO Document' (2009), pp. 11, 24–49. The arrow indicates the linearity of events, as suggested by the WHO's model (p. 24).

Bibliography

Abath, F.G.C., Montenegro, S.M.L. and Gomes, Y.M. (1998), 'Vaccines against Human Parasitic Diseases: An Overview', *Acta Tropica*, 71, 3, 237–254.
Abeysinghe, S. and White, K. (2010), 'Framing Disease: The Avian Influenza Pandemic in Australia', *Health Sociology Review*, 19, 3, 369–381.
Abeysinghe, S. and White, K. (2011), 'The Avian Influenza Pandemic: Discourses of Risk, Contagion and Preparation in Australia', *Health, Risk & Society*, 13, 4, 311–326.
Abraham, T. (2011), 'The Chronicle of a Disease Foretold: Pandemic H1N1 and the Construction of a Global Health Security Threat', *Political Studies*, 59, 4, 797–812.
Alaszewski, A. (2003), 'Risk, Trust and Health', *Health, Risk & Society*, 5, 3, 235–239.
Ali, S.H. (2009), 'SARS as an Emergent Complex: Towards a Networked Approach to Urban Infectious Disease', *Networked Disease*, 235–249.
Ali, S.H. and Keil, R. (2006), 'Global Cities and the Spread of Infectious Disease: The Case of Severe Acute Respiratory Syndrome (SARS) in Toronto, Canada', *Urban Studies*, 43, 491–509.
Arber, A. (1954), *The Mind and the Eye: The Study of the Biologist's Standpoint*, Cambridge, Cambridge University Press.
Aronowitz, R.A. (1991), 'Lyme Disease: The Social Construction of a New Disease and Its Social Consequences', *Milbank Quarterly*, 69, 1, 79–112.
Auge, M. and Herzlich, C. (1995), *The Meaning of Illness: Anthropology, History and Sociology*, Paris, Harwood Academic Press.
Autin, LJ. (1980), *How to Do Things with Words*, Oxford, Oxford University Press.
Baker, M. and Fidler, D.P. (2006), 'Global Public Health Surveillance under New International Health Regulations', *Emerging Infectious Diseases*, 12, 7, 1058–1655.
Bancroft, A. (2001), 'Globalisation and HIV/AIDS: Inequality and the Boundaries of a Symbolic Epidemic', *Health, Risk & Society*, 3, 1, 89–98.
Bancroft-Hinchey, T. (2010), 'H1N1, the False Pandemic', *Pravda*, http://english.pravda.ru/health/13-01-2010/111631-falsepandemic-0/ [accessed July 2010].
Barbour, R.S. and Huby, G. (1998), *Meddling with Mythology: AIDS and the Social Construction of Knowledge*, London, Routledge.
Barnes, B. (1974), *Scientific Knowledge and Sociological Theory*, London, Routledge and K. Paul.
Barry, J.M. (2004), *The Great Influenza: The Epic Story of the Deadliest Plague in History*, London, Penguin.
Bartholemew, D.J. (1995), 'What Is Statistics?' *Journal of the Royal Statistical Society*, 158, 1, 1–12.
Bashford, A. (2002), 'At the Border: Contagion, Immigration, Nation', *Australian Historical Studies*, 120, 345–348.
Bashford, A. and Strange, C. (2003), 'Isolation and Exclusion in the Modern World', *in* Strange, C. and Bashford, A. (eds.), *Isolation: Places and Practices of Exclusion*, London, Routledge.

Bauman, Z. (1998), *Globalization: The Human Consequences*, Cambridge, Polity Press.
Bauman, Z. (1999), *Post-Modernity and Its Discontents*, Cambridge, Polity Press.
Beard, R.L. (2004), 'Advocating Voice: Organisational, Historical and Social Milieux of the Alzheimer's Disease Movement', *Sociology of Health & Illness*, 26, 6, 797–819.
Beck, U. (1992), *Risk Society: Towards a New Modernity*, London, Sage.
Beck, U. (1999), *World Risk Society*, Cambridge, Polity Press.
Beigbeder, Y. (1998), *The World Health Organization*, The Hague, Martinus Nijhoff.
Best, J. (2001), *Damned Lies and Statistics: Untangling Numbers from the Media, Politicians, and Activists*, Berkeley, University of California.
Beverley, P.C.L. (2002), 'Immunology of Vaccination', *British Medical Bulletin*, 62, 1, 15–28.
Bloor, D. (1976), *Knowledge and Social Imagery*, London, Routledge & Keagan Paul.
Blyth, M. (2003), 'Structures Do Not Come with an Instruction Sheet: Interests, Ideas, and Progress in Political Science', *Perspectives on Politics*, 1, 4, 695–706.
Blyth, M. (2007), 'Powering, Puzzling, or Persuading? The Mechanisms of Building Institutional Orders', *International Studies Quarterly*, 51, 4, 761–777.
Bowker, G. and Star, S.L. (1991), 'Situations vs. Standards in Long-Term, Wide-Scale Decision-Making: The Case of the International Classification of Diseases', *Proceedings of the Twenty-Fourth Annual Hawaii International Conference on Social Systems*, Kauai, Hawaii.
Bowker, G. and Star, S.L. (1999), *Sorting Things Out: Classification and Its Consequences*, New Baskerville, Massachusetts Institute of Technology.
Braun, K. and Kropp, C. (2010), 'Beyond Speaking Truth? Institutional Responses to Uncertainty on Scientific Governance', *Science, Technology & Human Values*, 35, 771–782.
Brown, P. (1995), 'Naming and Framing: The Social Construction of Diagnosis and Illness', *Journal of Health and Social Behavior*, 35, 34–52.
Brown, P. (2000), 'Popular Epistemology and Toxic Waste Contamination: Lay and Professional Ways of Knowing', *in* Kroll-Smith, S., Brown, P. and Gunter, V.J. (eds.), *Illness and the Environment: A Reader in Contested Medicine*, New York, New York University Press, pp. 364–383.
Brown, T.M., Cueto, M. and Fee, E. (2006), 'The World Health Organization and the Transition from "International" to "Global" Public Health', *American Journal of Public Health*, 96, 1, 62–72.
Brown, B., Nerlich, B., Crawford, P., Koteyko, N. and Carter, R. (2009), 'Hygiene and Biosecurity: The Language of Politics and Risk in an era of Emerging Infectious Disease', *Sociology Compass*, 2, 6, 1–13.
Buse, K. and Walt, G. (2000), 'Global Private-Public Partnerships: Part II – What Are the Health Issues for Global Governance', *Bulletin of the World Health Organisation*, 78, 5, 699.
Callon, M. (1986), 'Some Elements of a Sociology of Translation: Domestication of the Scallops and the Fishermen of St Bruic Bay', *in* Law, J. (ed.), *Power, Action and Belief: A New Sociology of Knowledge?*, London, Routledge.
Callon, M. and Rabeharisoa, V. (2003), 'Research "in the Wild" and the Shaping of New Social Identities', *Technology in Society*, 25, 2, 193–204.

Campbell, J.L. (2010), 'Institutional Reproduction and Change', *in* Morgan, G., Campbell, J.L., Crouch, C., Pedersen, O.K. and Whitley, R. (eds.), *The Oxford Handbook of Comparative Institutional Change*, Oxford, Oxford University Press.
Cannell, J.J., Zasloff, M., Garland, C.F., Scragg, R. and Giovannucci, E. (2008), 'On the Epidemiology of Influenza', *Virology*, 5, 1, 29–39.
Caprara, A. (1998), 'Cultural Interpretations of Contagion', *Tropical Medicine & International Health*, 3, 12, 996–1001.
Cohen, S. and Enserink, M. (2009), 'After Delays, WHO Agrees: The 2009 H1N1 Pandemic Has Begun', *Science*, 324, 5934, 1496–1497.
Collins, H.M. (1983), 'The Sociology of Scientific Knowledge: Studies of Contemporary Science', *Annual Review of Sociology*, 9, 265–285.
Collins, H.M. and Pinch, T.J. (1982), *Frames of Meaning: The Social Construction of Extraordinary Science*, London, Routledge & Kegan Paul.
Condit, C.M., Achter, P.J., Lauer, I. and Sefcovic, E. (2002), 'The Changing Meanings of "Mutations": A Contextualised Study of Public Discourse', *Human Mutation*, 16, 69–72.
Condit, C.M., Dubriwny, T., Lynch, J. and Parrott, R. (2004), 'Lay People's Understanding of and Preference against the Word "Mutation"', *American Journal of Medical Genetics Part A*, 130A, 3, 245–250.
Corrigan, P. (1979), *The World Health Organization*, Hove (East Sussex), Wayland Publishers.
Council of Europe (2010), 'Council of Europe in Brief', http://www.coe.int/aboutCoe/index.asp?page=quisommesnous&l=en [accessed January 2010].
Cox, N.J. and Subbarao, K. (2000), 'Global Epidemiology of Influenza: Past and Present', *Annual Review of Medicine*, 51, 1, 407–421.
Crosby, A.W. (1976), *Epidemics and Peace*, Westport, Greenwood Press.
Da'dara, A.A. and Harn, D.A. (2005), 'DNA Vaccines against Tropical Parasitic Disease', *Expert Review of Vaccines*, 4, 4, 575–589.
Darmon, G. (1986), 'The Asymmetry of Symmetry', *Social Science Information*, 25, 3, 743–755.
David, P.A. (1994), 'Why Are Institutions the "Carriers of History"?: Path Dependence and the Evolution of Conventions, Organizations and Institutions', *Structural Change and Economic Dynamics*, 5, 2, 205–220.
Davis, M. (2011), 'Living "Post-Pandemic" and Responding to Influenza', *The Australian Sociological Association Conference,* November 28th–December 1st, 2011, Newcastle, TASA.
Declich, S. and Carter, A.O. (1994), 'Public Health Surveillance: Historical Origins, Methods and Evaluation', *Bulletin of the World Health Organization*, 72, 2, 285–304.
Derksen, L. (2000), 'Towards a Sociology of Measurement', *Social Studies of Science*, 30, 6, 803–845.
Dijk, T.A.v. (2001), 'Critical Discourse Analysis', *in* Schiffrin, D., Tannen, D. and Hamilton, H.E. (eds.), *The Handbook of Discourse Analysis*, Malden, Mass., Blackwell Publishers.
Djelic, M. (2010), 'Institutional Perspectives – Working Towards Coherence or Irreconcilable Diversity?' *in* Morgan, G., Campbell, J.L., Crouch, C., Pedersen, O.K. and Whitley, R. (eds.), *The Oxford Handbook of Comparative Institutional Analysis*, Oxford, Oxford University Press.

Donaldson, A., Lowe, P. and Ward, N. (2002), 'Virus-Crisis-Institutional Change: The Foot and Mouth Actor Network and the Governance of Rural Affairs in the UK', *Sociologia Ruralis*, 42, 3, 201–214.
Douglas, M. (1969), *Purity and Danger: An Analysis of Concepts of Pollution and Taboo*, London, Routledge & Kegan Paul.
Douglas, M. (1973), *Natural Symbols: Explorations in Cosmology*, London, Barrie and Jenkins.
Douglas, M. (1989), *How Institutions Think*, Syracuse, NY, Syracuse University Press.
Douglas, M. (1994), *Risk and Blame: Essays in Cultural Theory*, London, Routledge.
Douglas, M. and Wildavsky, A. (1983), *Risk and Culture: An Essay on the Selection of Technological and Environmental Dangers*, Berkeley and LA, University of California Press.
Dubos, R. and Dubos, G. (1953), *The White Plague: Tuberculosis, Man and Society*, London, Victor Gollancz Ltd.
Dudo, A.D., Dahlstrom, M.F. and Brossard, D. (2007), 'Reporting a Potential Pandemic: A Risk- Related Assessment of Avian Influenza Coverage in U.S. Newspapers', *Science Communication*, 28, 4, 429–454.
Dupre, J. (2006), 'Scientific Classification', *Theory, Culture & Society*, 23, 2–3, 30–32.
Durkheim, E. (1961), *The Elementary Forms of Religious Life*, New York, Free Press.
Durkheim, E. and Mauss, M. (1963), *Primitive Classification*, Chicago, University of Chicago Press.
Eichelberger, L. (2007), 'SARS and New York's Chinatown: The Politics of Risk and Blame During an Epidemic of Fear', *Social Science & Medicine*, 65, 1284–1295.
Elling, R.H. (1981), 'The Capitalist World-System and International Health', *Journal of Health Services*, 11, 1, 21–51.
Epstein, S. (1996), *Impure Science: AIDS, Activism, and the Politics of Knowledge*, Berkeley, University of California.
Evans, D., Cauchemez, S. and Hayden, F.G. (2009), ' "Prepandemic" Immunization for Novel Influenza Viruses, "Swine Flu" Vaccine, Guillain-Barre Syndrome, and the Detection of Rare Severe Adverse Events', *Journal of Infectious Diseases*, 200, 3, 321–328.
Farmer, P. (1999), *Infections and Inequalities: The Modern Plagues*, Berkeley, University of California Press.
Farmer, P., Walton, D. and Tarter, L. (2000), 'Infections and Inequalities', *Global Change & Human Health*, 1, 2, 94–109.
Fee, E., Cueto, M. and Brown, T.M. (2008), 'WHO at 60: Snapshots From Its First Six Decades', *American Journal of Public Health*, 98, 4, 630–633.
Ferguson, N.M., Cummings, D.A.T., Cauchemez, S., Fraser, C., Riley, S., Meeyai, A., Iamsirithaworn, S. and Burke, D.S. (2005), 'Strategies for Containing an Emerging Influenza Pandemic in Southeast Asia', *Nature*, 437, 7056, 209–214.
Ferguson, N.M., Cummings, D.A.T., Fraser, C., Cajka, J.C., Cooley, P.C. and Burke, D.S. (2006), 'Strategies for Mitigating an Influenza Pandemic', *Nature*, 442, 7101, 448–452.
Fidler, D.P. (2001), 'The Globalization of Public Health: The First 100 years of International Health Diplomacy', *Bulletin of the World Health Organization*, 79, 9, 842–849.

Fidler, D.P. (2004), 'Germs, Governance, and Global Public Health in the Wake of SARS', *Journal of Clinical Investigation*, 113, 6, 799–804.
Fleck, L. (1979), *Genesis and Development of a Scientific Fact*, Chicago, University of Chicago.
Foege, W.H. (1991), 'Plagues: Perception of Risk and Social Responses', in Mack, A. (ed.), *In the Time of Plague: The History and Social Consequences of Lethal Epidemic Disease*, New York, New York University Press.
Fogarty, A., Holland, K., Imison, M., Blood, R.W., Chapman, S. and Holding, S. (2011), 'Communicating Uncertainty – How Australian Television Reported H1N1 Risk in 2009: A Content Analysis', *BMC Public Health*, 11, 1, 181.
Foucault, M. (1970), *The Order of Things: An Archaeology of the Human Sciences*, New York, Routledge.
Foucault, M. (1988), *Technologies of the Self: A Seminar with Michel Foucault*, Amherst, University of Massachusetts Press.
Foucault, M. (ed.) (1991), *The Foucault Effect: Studies in Governmentality: With Two Lectures by and an Interview with Michel Foucault*, London, Harvester Wheatsheaf.
Francis, L.P., Battin, M.P., Jacobson, J.A., Smith, C.B. and Botkin, J. (2005), 'How Infectious Diseases Got Left Out – and What This Omission Might Have Meant for Bioethics', *Bioethics*, 19, 4, 307–322.
Franozi, R. (1998), 'Narrative Analysis – Or Why (and How) Sociologists Should be Interested in Narrative', *Annual Review of Sociology*, 24, 517–554.
Freeman, R. and Frisina, L. (2010), 'Health Care Systems and the Problem of Classification', *Journal of Comparative Policy Analysis: Research and Practice*, 12, 1, 163–178.
Friese, C. (2010), 'Classification Conundrums: Categorizing Chimeras and Enacting Species Preservation', *Theory and Society*, 39, 2, 145–172.
Funtowicz, S. and Ravetz, J. (1993), 'The Emergence of Post-Normal Science', in von Schomberg, R. (ed.), *Science, Politics and Morality: Scientific Uncertainty and Decision Making*, Dordrecht, Kluwer.
Funtowicz, S.O. and Ravetz, J.R. (1994), 'Uncertainty, Complexity and Post-Normal Science', *Environmental Toxicology and Chemistry*, 13, 12, 1881–1885.
Gandy, M. and Zumla, A. (2002), 'The Resurgence of Disease: Social and Historical Perspectives on the "New" Tuberculosis', *Social Science & Medicine*, 55, 3, 385–396.
Gensini (2004), 'The Concept of Quarantine in History: From Plague to SARS', *Journal of Infection*, 49, 257–261.
Gibson, N., Cave, A., Doering, D., Ortiz, L. and Harms, P. (2005), 'Socio-Cultural Factors Influencing Prevention and Treatment of Tuberculosis in Immigrant and Aboriginal Communities in Canada', *Social Science & Medicine*, 61, 5, 931–942.
Giddens, A. (1991), *Modernity and Self-Identity: Self and Society in the Late Modern Age*, Cambridge, Polity.
Giddens, A. (1999), 'Risk and Responsibility', *Modern Law Review*, 62, 1, 1–10.
Gieryn, T.F. (1983), 'Boundary-Work and the Demarcation of Science from Non-Science: Strains and Interests in Professional Ideologies of Scientists', *American Sociological Review*, 48, 6, 781–795.
Gigerenzer, G. (2002), *Reckoning with Risk*, London, Penguin Books.

Gollier, C. and Treich, N. (2003), 'Decision-Making under Scientific Uncertainty: The Economics of the Precautionary Principle', *Journal of Risk and Uncertainty*, 27, 1, 77–103.
Gorges, M.J. (2001), 'New Institutionalist Explanations for Institutional Change: A Note of Caution', *Politics*, 21, 2, 137–145.
Greenwald, H.P. (2008), *Organizations: Management without Control*, Thousand Oaks, Sage.
Greif, A. and Laitin, D.D. (2004), 'A Theory of Endogenous Institutional Change', *American Political Science Review*, 98, 4, 633–652.
Grundmann, R. (2006), 'Ozone and Climate: Scientific Consensus and Leadership', *Science, Technology & Human Values*, 31, 1, 73–101.
Haber, P., DeStefano, F., Angulo, F.J., Iskander, J., Shadomy, S.V., Weintraub, E. and Chen, R.T. (2004), 'Guillain-Barre Syndrome Following Influenza Vaccination', *JAMA: The Journal of the American Medical Association*, 292, 20, 2478–2481.
Hacking, I. (1999), *The Taming of Chance*, Cambridge, Cambridge University Press.
Hall, P.A. (2010), 'Historical Institutionalism in Rationalist and Sociological Perspective', *in* Mahoney, J. and Thelen, K. (eds.), *Explaining Institutional Change*, Cambridge, Cambridge University Press.
Hall, P.A. and Taylor, R.C.R. (1996), 'Political Science and the Three New Institutionalisms', *Political Studies*, 44, 5, 936–957.
Hallowell, N. (1999), 'Doing the Right Thing: Genetic Risk and Responsibility', *Sociology of Health and Illness*, 21, 5, 597–621.
Hansson, S.O. (2005), 'Seven Myths of Risk', *Risk Management*, 7, 2, 7–17.
Haraway, D. (1991), *Simians, Cyborgs and Women: The Reinvention of Nature*, New York, Routledge.
Hay, C. and Wincott, D. (1998), 'Structure, Agency and Historical Institutionalism', *Political Studies*, 46, 5, 951–957.
Hayden, F.G. (2006), 'Antiviral Resistance in Influenza Viruses: Implications for Management and Pandemic Response', *New England Journal of Medicine*, 354, 8, 785–788.
Herzlich, C. and Pierret, J. (1987), *Illness and Self in Society*, London, John Hopkins University Press.
Hindess, B. (1973), *The Use of Official Statistics in Sociology: A Critique of Positivism and Ethnomethodology*, London, McMillan Press.
Hooker, C. and Ali, S.H. (2009), 'SARS and Security: Health in the "New Normal"', *Studies in Political Economy*, 84, 101–126.
Hughes, R.A.C., Hadden, R.D.M., Gregson, N.A. and Smith, K.J. (1999), 'Pathogenesis of Guillain-Barré Syndrome', *Journal of Neuroimmunology*, 100, 1–2, 74–97.
Inhorn, M.C. and Brown, P.J. (1990), 'The Anthropology of Infectious Disease', *Annual Review of Anthropology*, 19, 1, 89–117.
Janes, C.R. and Corbett, K.K. (2009), 'Anthropology and Global Health', *Annual Review of Anthropology*, 38, 167–183.
Jasanoff, S. (2004a), 'The Idiom of Co-Production', *in* Jasanoff, S. (ed.), *States of Knowledge: The Co-Production of Science and Social Order*, London, Routledge.
Jasanoff, S. (2004b), 'Ordering Knowledge, Ordering Society', *in* Jasanoff, S. (ed.), *States of Knowledge: The Co-Production of Science and Social Order*, London, Routledge.

Jasanoff, S.S. (1987), 'Contested Boundaries in Policy-Relevant Science', *Social Studies of Science*, 17, 2, 195–230.
Kavanagh, A.M. and Broom, D.H. (1998), 'Embodied Risk: My Body, Myself?' *Social Science & Medicine*, 46, 3, 437–444.
Karkazis, K and Feder, E.K. (2008), 'Naming the Problem: Disorders and Their Meanings', *Lancet*, 372, 9655, 216–2017.
Keane, C. (1998), 'Globality and Constructions of World Health', *Medical Anthropology Quarterly*, 12, 2, 226–240.
Keil, R. and Ali, H. (2007), 'Governing the Sick City: Urban Governance in the Age of Emerging Infectious Disease', *Antipode*, 39, 5, 846–873.
Kickbusch, I. and de Leeuw, E. (1999), 'Global Public Health: Revisiting Healthy Public Policy at the Global Level', *Health Promotion International*, 14, 4, 285–288.
King, N.B. (2002), 'Security, Disease, Commerce: Ideologies of Postcolonial Global Health', *Social Studies of Science*, 32, 5–6, 763–789.
Kleinman, D.L. and Kinchy, A.J. (2003), 'Boundaries in Science Policy Making', *Sociological Quarterly*, 44, 4, 577–595.
Krippendorf, K. (1980), *Content Analysis: An Introduction to Its Methodology*, London, SAGE Publications.
Kuhn, T.S. (1970), *The Structure of Scientific Revolutions*, Chicago, University of Chicago Press.
Kushner, H.I., Turner, C.L., Bastian, J.F. and Burns, J.C. (2004), 'The Narratives of Kawasaki Disease', *Bulletin of the History of Medicine*, 78, 2, 410–439.
Laet, M. and Mol, A. (2000), 'The Zimbabwe Bush Pump: Mechanics of a Fluid Technology', *Social Studies of Science*, 30, 2, 225–263.
Lamont, M. and Molnar, V. (2002), 'The Study of Boundaries in the Social Sciences', *Annual Review of Sociology*, 28, 167–195.
Langmuir, A.D. (1979), 'Guillain-Barre Syndrome: The Swine Influenza Virus Vaccine Incident in the United States of America, 1976–77: Preliminary Communication', *Journal of the Royal Society of Medicine*, 72, 9, 660–669.
Lash, S. (2005), 'Risk Culture', *in* Adam, B., Beck, U. and Van Loon, J. (eds.), *The Risk Society and Beyond*, London, Sage.
Latour, B. (1987), *Science in Action: How to Follow Scientists and Engineers through Society*, Cambridge, MA, Harvard University Press.
Latour, B. (1996), 'On Actor-Network Theory: A Few Clarifications Plus More than a Few Complications', *Soziale Welt*, 47, 4, 369–381.
Latour, B. (1998), 'Essays on Science and Society: From the World of Science to the World of Research?' *Science*, 280, 5361, 208–209.
Latour, B. (2004), 'Why Has Critique Run Out of Steam? From Matters of Fact to Matters of Concern', *Critical Inquiry*, 30, 225–248.
Latour, B. (2005), *Reassembling the Social: An Introduction to Actor-Network Theory*, Oxford, Oxford University Press.
Latour, B. and under alias of Johnson, J. (1988), 'Mixing Humans and Nonhumans Together: The Sociology of the Door-Closer', *Social Problems*, 35, 3, 298–310.
Latour, B. and Woolgar, S. (1979), *Laboratory Life: The Social Construction of Scientific Facts*, Beverley Hills, Sage Publications.
Law, J. (1992), 'Notes on the Theory of the Actor-Network: Ordering, Strategy, and Heterogeneity', *Systemic Practice and Action Research*, 5, 4, 379–393.

Law, J. (2006), 'Disaster in Agriculture, or Foot and Mouth Mobilities', *Environment and Planning A*, 38, 133–143.
Law, J. and Singleton, V. (2009), 'A Further Species of Trouble?' *in* Doering, M. and Nerlich, B. (eds.), *From Mayhem to Meaning: The Cultural Meaning of the 2001 Outbreak of Foot and Mouth Disease in the UK*, Manchester, Manchester Press.
Lazzari, S. and Stohr, K. (2004), 'Avian Influenza and Influenza Pandemics', *Bulletin of the World Health Organisation*, 82, 4, 242–242A.
Lee, K. (2003), 'Introduction', *in* Lee, K. (ed.), *Health Impacts of Globalization: Towards Global Governance*, Basingstoke, Palgrave Macmillan.
Lee, K. (2005), *Globalisation and Health: An Introduction*, New York, Palgrave McMillan.
Lenhard, J., Lucking, H. and Schwechheimer, H. (2006), 'Expert Knowledge, Mode-2 and Scientific Disciplines: Two Contrasting Views', *Science and Public Policy*, 33, 5, 341–350.
Levidow, L. (2001), 'Precautionary Uncertainty: Regulating GM Crops in Europe', *Social Studies of Science*, 31, 842–874.
Lewin, P. (1994), 'Categorization and the Narrative Structure of Science', *Philosophy & Rhetoric*, 27, 1, 35–62.
Liess, W. and Hrudey, S.E. (2003), 'Risk Management and Precaution: Insights in the Cautious Use of Evidence', *Environmental Health Perspective*, 111, 13, 1577–1581.
Lipsitch, M., Cohen, T., Murray, M. and Levin, B.R. (2007), 'Antiviral Resistance and the Control of Pandemic Influenza', *PLoS Med*, 4, 1, e15.
Lohm, D. (2011), ' "Soldier On" ' or Surrender? Military Discourse and Its Dilemmas for Australians Responding to Influenza', *The Australian Sociological Association Conference*, November 28th–December 1st, 2011, Newcastle, TASA.
Lowndes, V. (2002), 'Institutionalism', *in* Marsh, D. and Stroker, G. (eds.), *Theory and Methods in Political Science* (2 Edition), Basingstoke, Palgrave Macmillan.
Lowndes, V. (2010), 'The Institutional Approach', *in* Marsh, D. and Stroker, G. (eds.), *Theory and Methods in Political Science* (3 Edition), Basingstoke, Palgrave Macmillan.
Luhmann, N. (2002), *Risk: A Sociological Theory*, New Brunswick, NJ, Transaction Publishers.
Lupton, D. (1994), *Moral Threats and Dangerous Desires: AIDS and the News Media*, London, Taylor and Francis.
Lupton, D. (1999), *Risk*, London, Routledge.
Lynch, M. (2004), 'Circumscribing Expertise: Membership Categories in Courtroom Testimony', *in* Jasanoff, S. (ed.), *States of Knowledge: The Co-Production of Science and Social Order*, London, Routledge.
MacGregor, C. (2011), 'Hope in the Time of Uncertainty: Public Engagement with Science, Ontological Security and the H1N1 Influenza Pandemic in Australia', *The Australian Sociological Association Conference*, November 28th–December 1st, 2011, Newcastle, TASA.
Mack, E. (2006/2007), 'The World Health Organization's New International Health Regulations: Incursions on State Sovereignty and Ill-Fated Response to Global Health Issues', *Chicago Journal of International Law*, 7, 1, 365–377.
MacKenzie, D. (1996), *Knowing Machines: Essays on Technical Change*, Cambridge, MIT Press.

MacKenzie, D. and Wajcman, J. (1999), *The Social Shaping of Technology*, Buckingham, Open University Press.
Maguire, S. and Hardy, C. (2006), 'The Emergence of New Global Institutions: A Discursive Perspective', *Organization Studies*, 27, 1, 7–29.
Mahajan, M. (2008), 'Designing Epidemics: Models, Policy-Making, and Global Foreknowledge in India's AIDS Epidemic', *Science and Public Policy*, 35, 585–596.
Mahoney, J. (2000), 'Path Dependence in Historical Sociology', *Theory and Society*, 29, 507–548.
Mahoney, J. and Thelen, K. (2010), 'A Theory of Gradual Institutional Change', *in* Mahoney, J. and Tannen, D. (eds.), *Explaining Institutional Change*, Cambridge, Cambridge University Press.
Marková, I. and Farr, R. (1995), *Representations of Health, Illness and Handicap*, Chur, Harvard Academic Publishers.
Marshall, B.K. and Picou, J.S. (2008), 'Postnormal Science, Precautionary Principle, and Worst Cases: The Challenge of Twenty-First Century Catastrophes', *Sociological Inquiry*, 78, 2, 230–247.
Martin, A. (2004), 'Can't Any Body Count? Counting as an Epistemic Theme in the History of Human Chromosomes', *Social Studies of Science*, 34, 6, 923–948.
Martin, A. and Lynch, M. (2009), 'Counting Things and People: The Practices and Politics of Counting', *Social Problems*, 56, 2, 243–266.
Martinez, L. (2000), 'Global Infectious Disease Surveillance', *International Journal of Infectious Diseases*, 4, 4, 222–228.
Mathews, J.D., Chesson, J.M., McCaw, J.M. and McVernon, J. (2009), 'Understanding Influenza Transmission, Immunity and Pandemic Threats', *Influenza and Other Respiratory Viruses*, 3, 4, 143–149.
McCaw, J.M. and McVernon, J. (2007), 'Prophylaxis or Treatment? Optimal Use of an Antiviral Stockpile during an Influenza Pandemic', *Mathematical Biosciences*, 209, 2, 336–360.
McMichael, A.J. (2001), 'Human Culture, Ecological Change, and Infectious Disease', *Ecosystem Health*, 7, 2, 107–115.
McNeill, W.H. (1976), *Plagues and Peoples*, Harmondsworth, Penguin Books.
Merton, R.K. (1968), *Social Theory and Social Structure*, New York, Free Press.
Merton, R.K. (1973), *The Sociology of Science: Theoretical and Empirical Investigations*, Chicago, University of Chicago Press.
Miller, C.A. (2004), 'Climate Science and the Making of a Global Public Order', *in* Jasanoff, S. (ed.), *States of Knowledge: The Co-Production of Science and Social Order*, London, Routledge.
Monto, A.S. (1987), 'Influenza: Quantifying Morbidity and Mortality', *The American Journal of Medicine*, 82, 6, Supplement 1, 20–25.
Morse, S.S. (1995), 'Factors in the Emergence of Infectious Diseases', *Emerging Infectious Diseases*, 1, 1, 7–15.
Moscovici, S. (1988), 'Notes Towards a Description of Social Representations', *Journal of European Social Psychology*, 18, 211–250.
Moynihan, R. (2002), 'Disease-Mongering: How Doctors, Drug Companies, and Insurers are Making you Sick', *British Medical Journal*, 324, 923.
Moynihan, R., Health, I. and Henry, D. (2002), 'Selling Sickness: The Pharmaceutical Industry and Disease Mongering', *British Medical Journal*, 324, 886–891.

Murdoch, J. (2001), 'Ecologising Sociology: Actor-Network Theory, Co-Construction and the Problem of Human Exemptionalism', *Sociology*, 35, 1, 111–133.
Mykhalovskiy, E. and Weir, L. (2006), 'The Global Public Health Intelligence Network and Early Warning Outbreak Detection: A Canadian Contribution to Public Health', *Canadian Journal of Public Health*, 97, 1, 42–44.
Mythen, G. (2007), 'Reappraising Risk Society: Telescopic Sight or Myopic Vision', *Current Sociology*, 55, 793–813.
Navarro, V. (1983), 'Radicalism, Marxism, and Medicine', *International Journal of Health Services*, 13, 2, 179–202.
Navarro, V. (1985), 'US Marxist Scholarship in the Analysis of Health and Medicine', *International Journal of Health Services*, 15, 4, 525–545.
Ncayiyana, D.J. (2010), 'H1N1 Hype a Treasure Trove for Conspiracy Theorists', *South African Medical Journal*, 100, 131–131.
Nelkin, D. and Gilman, S. (1991), 'Placing Blame for Devastating Disease', *in* Mack, A. (ed.), *In Time of Plague: The History and Social Consequences of Lethal Epidemic Disease*, New York, New York University Press.
Nerlich, B. and Halliday, C. (2007), 'Avian Flu: The Creation of Expectations in the Interplay between Science and the Media', *Sociology of Health & Illness*, 29, 1, 46–65.
Nerlich, B. and Koteyko, N. (2011), 'Crying Wolf? Biosecurity and Metacommunication in the Context of the 2009 Swine Flu Pandemic', *Health & Place*, 18, 4, 710–717.
Nguyen-Van-Tam, J.S. and Hampson, A.W. (2003), 'The Epidemiology and Clinical Impact of Pandemic Influenza', *Vaccine*, 21, 16, 1762–1768.
Nowotny, H. (2003a), 'Democratising Expertise and Socially Robust Knowledge', *Science and Public Policy*, 30, 3, 151–156.
Nowotny, H. (2003b), 'Re-Thinking Science: From Reliable Knowledge to Socially Robust Knowledge', *in* Lepenies, W. (ed.), *Entangled Histories and Negotiated Universals: Centres and Peripheries in a Changing World*, Fankfurt, Campus.
Nowotny, H., Scott, P. and Gibbons, M. (2001), *Re-Thinking Science: Knowledge and the Public in an Age of Uncertainty*, Malden, Polity Press.
Odent, B. (2010), 'Influenza A: "They Organized a Psychosis" ', *Humanité (English Translation Version)*, http://www.humaniteinenglish.com/spip.php?article1427.
Ollila, E. (2005), 'Global Health Priorities – Priorities of the Wealthy?' *Globalization and Health*, 1, 6, 6–11.
Patel, A. and Gorman, S.E. (2009), 'Stockpiling Antiviral Drugs for the Next Influenza Pandemic', *Clinical Pharmacology and Therapeutics*, 86, 3, 241–243.
Pels, D. (1996), 'The Politics of Symmetry', *Social Studies of Science*, 26, 2, 277–304.
Petersen, A. (1996), *The New Public Health: Health and Self in the Age of Risk*, London, Sage.
Pierson, P. (2000), 'The Limits of Design: Explaining Institutional Origins and Change', *Governance*, 13, 4, 475–499.
Pinch, T.J. and Bijker, W.E. (1984), 'The Social Construction of Facts and Artefacts: Or How the Sociology of Science and the Sociology of Technology Might Benefit Each Other', *Social Studies of Science*, 14, 3, 399–441.
Pohl, C. (2008), 'From Science to Policy through Transdisciplinary Research', *Environmental Science & Policy*, 11, 1, 46–53.

Popper, K. (1959), *The Logic of Scientific Discovery*, London, Hutchinson.
Porta, M. (2008), *A Dictionary of Epidemiology – 5th Edition*, Oxford, Oxford University Press.
Prout, A. (1996), 'Actor-Network Theory, Technology and Medical Sociology: An Illustrative Analysis of the Metered Dose Inhaler', *Sociology of Health & Illness*, 18, 2, 198–219.
Quah, S.R. (2007), 'Public Image and Governance of Epidemics: Comparing HIV/AIDS and SARS', *Health Policy*, 80, 253–272.
Ravetz, J. (2004), 'The Post-Normal Science of Precaution', *Futures*, 36, 3, 347–357.
Reich, M.R. (2000), 'Private-Public Partnerships for Public Health', *Nature Medicine*, 6, 6, 216–620.
Reinicke, W.H. (1999), 'The Other World Wide Web: Global Public Policy Networks', *Foreign Policy*, 117, 44–57.
Robertson, R. (1995). 'Glocalization: Time-Space and Homogeneity-Heterogeneity', *in* Featherstone, M., Lash. S. and Robertson. R. (eds.), *Global Modernities*, pp. 25–44. London, Sage.
Rose, N., O'Malley, P. and Valverde, M. (2006), 'Governmentality', *Annual Review of Law and Social Science*, 2, 1, 83–104.
Rose, S.P. and Rose, H. (1976), *The Political Economy of Science: Ideology of/in the Natural Sciences*, London, Macmillan.
Rosebury, T. (1971), *Microbes and Morals: The Strange History of Venereal Disease*, London, Secker & Warburg.
Rothstein, H. (2006), 'The Institutional Origins of Risk: A New Agenda for Risk Research', *Health, Risk & Society*, 8, 3, 215–221.
Rothstein, H., Huber, M. and Gaskell, G. (2006), 'A Theory of Risk Colonization: The Spiraling Regulatory Logics of Societal and Institutional Risk', *Economy and Society*, 35, 1, 91–112.
Saloranta, T. (2001), 'Post-Normal Science and the Global Climate Change Issue', *Climatic Change*, 50, 4, 395–404.
Schmidt, V.A. (2008), 'Discursive Institutionalism: The Explanatory Power of Ideas and Discourse', *Annual Review of Political Science*, 11, 303–326.
Schmidt, V.A. (2010), 'Taking Ideas and Discourse Seriously: Explaining Change Through Discursive Institutionalism as the Fourth "New Institutionalism"', *European Political Science Review*, 2, 1, 1–25.
Selgelid, M.J. (2005), 'Ethics and Infectious Disease', *Bioethics*, 19, 3, 272–289.
Shackley, S. and Wynne, B. (1996), 'Representing Uncertainty in Global Climate Change Science and Policy: Boundary-Ordering Devices and Authority', *Science, Technology & Human Values*, 21, 3, 275–302.
Shapin, S. (1994), *A Social History of Truth: Civility and Science in Seventeenth Century England*, Chicago, University of Chicago Press.
Shapin, S. (1998), *Science Incarnate: Historical Embodiments of Natural Knowledge*, Chicago, University of Chicago Press.
Shrader-Frechette, K. (1993), 'Probabilistic Uncertainty and Technological Risks', *in* von Schomberg, R. (ed.), *Science, Politics and Morality: Scientific Uncertainty and Decision Making*, Dordrecht, Kluwer.
Silverman, D. (2004), *Interpreting Qualitative Data: Methods for Analysing Talk, Text and Interaction*, London, Sage.
Silverman, M. (1976), *The Drugging of the Americas*, Berkeley, University of California Press.

Singer, M.C., Erickson, P.I., Badiane, L., Diaz, R., Ortiz, D., Abraham, T. and Nicolaysen, A.M. (2006), 'Syndemics, Sex and the City: Understanding Sexually Transmitted Diseases in Social and Cultural Context', *Social Science & Medicine*, 63, 8, 2010–2021.

Smith, R.D. (2006), 'Responding to Global Infectious Disease Outbreaks: Lessons from SARS on the Role of Risk Perception, Communication and Management', *Social Science & Medicine*, 63, 12, 3113–3123.

Snowden, F.M. (2008), 'Emerging and Reemerging Diseases: A Historical Perspective', *Immunological Reviews*, 225, 9–26.

Sontag, S. (1978), *Illness as Metaphor*, Harmondsworth, Penguin.

Sontag, S. (1989), *Aids and Its Metaphors*, New York, Farrar Straus Giroux.

Stebbing, M. (2009), 'Avoiding the Trust Deficit: Public Engagement, Values, the Precautionary Principle and the Future of Nanotechnology', *Journal of Bioethical Inquiry*, 6, 1, 37–48.

Stephenson, N. (2011), 'The Splintering of Public Health's Public in Pandemic Preparedness Efforts: How Human Rights Approaches Work with the Securitisation of Health', *The Australian Sociological Association Conference*, November 28th–December 1st, 2011, Newcastle, TASA.

Susser, M. and Susser, E. (1996), 'Choosing a Future for Epidemiology: I. Eras and Paradigms', *American Journal of Public Health*, 86, 5, 668–673.

Swistak, P. (1990), 'Paradigms of Measurement', *Theory and Decision*, 29, 1, 1–17.

Szlezak, N., Bloom, A., Jamison, D.T., Keusch, G.T., Michaud, C.M., Moon, S. and Clark, W.C. (2010), 'The Global Health System: Actors, Norms and Expectations in Transition', *PLoS Medicine*, 7, 1, 1–4.

Taubenberger, J.K. and Morens, D.M. (2006), '1918 Influenza: The Mother of All Pandemics', *Emerging Infectious Diseases*, 12, 1, 15–22.

Tausczik, Y., Faasse, K., Pennebaker, J.W. and Petrie, K.J. (2012), 'Public Anxiety and Information Seeking Following the H1N1 Outbreak: Blogs, Newspaper Articles, and Wikipedia Visits', *Health Communication*, 27, 2, 179–185.

Tausig, M., Selgelid, M.J., Subedi, S. and Subedi, J. (2006), 'Taking Sociology Seriously: A New Approach to the Bioethical Problems of Infectious Disease', *Sociology of Health & Illness*, 28, 6, 838–849.

Taylor, A.L. (2005), 'Governing the Globalization of Public Health', *Journal of Law, Medicine & Ethics*, Fall 2004, 500–508.

Tognotti, E. (2003), 'Scientific Triumphalism and Learning from the Facts: Bacteriology and the 'Spanish Flu' Challenge of 1918', *The Journal of the Society for the Social History of Medicine*, 16, 1, 97–110.

Treacher, A. and Wright, P. (1982), *The Problem of Medical Knowledge: Examining the Social Construction of Medicine*, Edinburgh, Edinburgh University Press.

Turnbull, D. (1989), 'The Push for a Malaria Vaccine', *Social Studies of Science*, 19, 2, 283–300.

Vaihinger, H. (1949), *The Philosophy of 'As If': A System of the Theoretical, Practical and Religious Fictions on Mankind*. London, Routledge & Kegan Paul.

van Dijk, T.A. (2001), 'Critical Discourse Analysis', in Schiffrin, D., Tannen, D. and Hamilton, H.E. (eds.), *The Handbook of Discourse Analysis*, Malden, Blackwell Publishers.

van Loon, J. (2005), 'Epidemic Space', *Critical Public Health*, 15, 1, 39–52.

Vance, M.A. and Millington, W.R. (1986), 'Principles of Irrational Drug Therapy', *Journal of Health Services*, 16, 3, 355–362.

von Schomberg, R. (1993a), 'Controversies and Political Decision Making', in von Schomberg, R. (ed.), *Science, Politics and Morality: Scientific Uncertainty and Decision Making*, Dordrecht, Kluwer.
von Schomberg, R. (1993b), 'Introduction', in von Schomberg, R. (ed.), *Science, Politics and Morality: Scientific Uncertainty and Decision Making*, Dordrecht, Kluwer.
Wagner-Egger, P., Bangerter, A., Gilles, I., Green, E., Rigaud, D., Krings, F., Staerkla, C. and Clamence, A. (2011), 'Lay Perceptions of Collectives at the Outbreak of the H1N1 Epidemic: Heroes, Villains and Victims', *Public Understanding of Science*, 20, 4, 461–476.
Waitzkin, H. (1978), 'A Marxist View of Medical Care', *Annals of Internal Medicine*, 89, 2, 264–278.
Wald, P. (2008), *Contagious: Cultures, Carriers, and the Outbreak Narrative*, Durham and London, Duke University Press.
Walt, G. (1988), 'Globalisation of International Health', *The Lancet*, 351, 434–437.
Warren, A., Bell, M. and Budd, L. (2010), 'Airports, Localities and Disease: Representations of Global Travel During the H1N1 Pandemic', *Health & Place*, 16, 4, 727–735.
Washer, P. (2004), 'Representations of SARS in the British Newspapers', *Social Science & Medicine*, 59, 2561–2571.
Watts, S. (2003), *Disease and Medicine in World History*, New York, Routledge.
Webby, R.J. and Webster, R.G. (2003), 'Are We Ready for Pandemic Influenza', *Science*, 302, 5660, 1519–1522.
Weber, M. ([1913]1978), *Economy and Society: An Outline of Interpretive Sociology, Vol 2*, Berkeley, University of California Press.
Webster, R.G. (1997), 'Predictions for Future Human Influenza Pandemic', *Journal of Infectious Diseases*, 176, 514–519.
Weingart, P. (1999), 'Scientific Expertise and Political Accountability: Paradoxes of Science in Politics', *Science and Public Policy*, 26, 151–161.
Weir, L. and Mykhalovskiy, E. (2010), *Global Public Health Vigilance: Creating a World on Alert*, New York, Routledge.
Whittle, A. and Spicer, A. (2008), 'Is Actor Network Theory Critique?' *Organization Studies*, 29, 4, 611–629.
WHO (2007), *Working for Health: An Introduction to the World Health Organization*, Genveva, World Health Organization.
WHO (2011a), *Global Influenza Programme*, Geneva, World Health Organisation, http://www.who.int/influenza/en/ [accessed June 2010].
WHO (2011b), *Weekly Seasonal Influenza Updates*, Geneva, World Health Organisation, http://www.who.int/influenza/surveillance_monitoring/updates/GIP_surveillance_2011_archives/en/ [accessed November 2011].
Williams, S.J., Gabe, J. and Davis, P. (2008), 'The Sociology of Pharmaceuticals: Progress and Prospects', *Sociology of Health & Illness*, 30, 6, 813–824.
Willis, P. and Nerlich, B. (2005), 'Disease Metaphors in New Epidemics: The UK Media Framing of the 2003 SARS Epidemic', *Social Science & Medicine*, 60, 2629–2639.
Wittgenstein, L. ([1949] 1980), *Remarks on the Philosophy of Psychology*, Oxford, Blackwell.
Wittgenstein, L. ([1956] 1963), *Remarks on the Foundations of Mathematics*, Oxford, Blackwell.

Wodarg, W. and Villesen, K. (2009), 'Faked Pandemic', *Information (English Translation)*, http://www.wodarg.de/english/3013320.html [accessed July 2010].
Woodward, D. and Smith, R.D. (2003), 'Global Public Goods and Health: Health Economic and Public Health Perspectives', in Smith, R., Beaglehole, R., Woodward, D. and Drager, N. (eds.), *Global Public Boods for Health: Health Economic and Public Health Perspectives*, New York, Oxford University Press.
Yach, D. and Bettcher, D. (1998), 'The Globalization of Public Health, I: Threats and Opportunities', *American Journal of Public Health*, 88, 5, 735–738.
Yach, D. and Bettcher, D. (1998b), 'The Globalization of Public Health, II: The Convergence of Self-Interest and Altruism', *American Journal of Public Health*, 88, 5, 738–741.
Zinsser, H. (1942), *Rats, Lice, and History*, London, George Routledge & Sons.

Data Sources

WHO Situation Updates[1]

WHO (24/04/09–06/09/10), *Situation Updates – Pandemic (H1N1) 2009*. Update 1–112, available at: http://www.who.int/csr/disease/swineflu/updates/en/index.html

WHO Speeches and Statements

Chan, M. (25/04/09), *Swine Influenza 25/04/09*, Statement by WHO Director-General, available at: http://www.who.int/mediacentre/news/statements/2009/h1n1_20090425/en/index.html.
Chan, M. (27/04/09), *Swine Influenza 27/04/09*, Statement by WHO Director-General, available at: http://www.who.int/mediacentre/news/statements/2009/h1n1_20090427/en/index.html.
Chan, M. (29/04/09), *Influenza A(H1N1)*, Statement by WHO Director-General, available at: http://www.who.int/mediacentre/news/statements/2009/h1n1_20090429/en/index.html.
Chan, M. (04/05/09), *H1N1 Influenza Situation*, Statement made at the Secretary-General's Briefing to the United Nations General Assembly in the H1N1 Influenza Situation, available at: http://www.who.int/dg/speeches/2009/influenza_a_h1n1_situation_20090504/en/index.html.
Chan, M. (08/05/09), *World is Better Prepared for Influenza Pandemic*, Address to the ASEAN+3 Health Ministers' Special Meeting on Influenza (A)H1N1, Bangkok, Thailand, available at: http://www.who.int/dg/speeches/2009/asean_influenza_ah1n1_20090508/en/index.html.

---

[1] The Situation Updates are very short pieces of text, illustrating the epidemiology and geographic spread of H1N1 (112 in total). They were not used for in-depth textual analysis, though they informed the understanding of the WHO's perception of risk. The information from the Situation Updates were collated for Graph 1. They are not cited separately here, though all are accessible through the site indicated.

Chan, M. (15/05/09), *Sharing of Influenza Viruses, Access to Vaccines and Other Benefits*, Opening Remarks at the Intergovernmental Meeting on Pandemic Influenza Preparedness, Geneva, Switzerland, available at: http://www.who.int/dg/speeches/2009/pandemic_influenza_preparedness_20090515/en/index.html.

Chan, M. (18/05/09), *Concern Over Flu Pandemic Justified*, Address to the Sixty-Second World Health Assembly, Geneva, Switzerland, available at: http://www.who.int/dg/speeches/2009/62nd_assembly_address_20090518/en/index.html.

Chan, M. (18/05/09b), *Pandemic Threat Deserves Attention of All*, Remarks at the High-Level Consultation on Pandemic Influenza A (H1N1), Geneva, Switzerland, available at: http://www.who.int/dg/speeches/2009/influenza_h1n1_consultation_20090518/en/index.html.

Chan, M. (11/06/09), *World Now at the Start of 2009 Influenza Pandemic*, Statement to the Press by WHO Director-General, available at: http://www.who.int/mediacentre/news/statements/2009/h1n1_pandemic_phase6_20090611/en/index.html.

Chan, M. (17/06/09), *WHO Welcomes Sanofi-Aventis's Donation of Vaccine*, Statement by the WHO Director-General, available at: http://www.who.int/mediacentre/news/statements/2009/vaccine_donation_20090617/en/index.html.

Chan, M. (02/07/09), *Lessons Learned and Preparedness*, Keynote Speech at a High-Level Meeting on Influenza A(H1N1), Cancin, Quinta Roo, Mexico, available at: http://www.who.int/dg/speeches/2009/influenza_h1n1_lessons_20090702/en/index.html.

Chan, M. (18/09/09), *Pandemic Vaccine Donations for the Developing World*, Statement by the WHO Director-General, available at: http://www.who.int/mediacentre/news/statements/2009/pandemic_vaccine_donations_20090918/en/index.html.

Chan, M. (24/09/09), *Statement Following the Fifth Meeting of the Emergency Committee*, Statement by the WHO Director-General, available at: http://www.who.int/csr/disease/swineflu/5th_meeting_ihr/en/index.html.

Chan, M. (10/11/09), *Agreement for Donation of Pandemic H1N1 Vaccine Signed*, Statement by the WHO Director-General, available at: http://www.who.int/mediacentre/news/statements/2009/pandemic_vaccine_agreement_20091110/en/index.html.

Chan, M. (26/11/09), *Statement Following the Sixth Meeting of the Emergency Committee*, Statement by the WHO Director-General, available at: http://www.who.int/csr/disease/swineflu/6th_meeting_ihr/en/index.html.

Chan, M. (22/01/10), *Statement of the World Health Organization on Allegation of Conflict of Interest and 'Fake' Pandemic*, available at: http://www.who.int/mediacentre/news/statements/2010/h1n1_pandemic_20100122/en/index.html.

Chan, M. (24/02/10), *Director-General Statement Following the Seventh Meeting of the Emergency Committee*, available at: http://www.who.int/csr/disease/swineflu/7th_meeting_ihr/en/index.html.

Chan, M. (03/06/10), *Director-General Statement Following the Eighth Meeting of the Emergency Committee*, available at: http://www.who.int/csr/disease/swineflu/8th_meeting_ihr/en/index.html.

230 Bibliography

Chan, M. (08/06/10) *WHO Director-General's Letter to BMJ Editors*, available at: http://www.who.int/mediacentre/news/statements/2010/letter_bmj_20100608/en/index.html.
Chan, M. (10/08/10), *H1N1 in Post-Pandemic Period: Director-General's Statement at Virtual Press Conference*, available at: http://www.who.int/mediacentre/news/statements/2010/h1n1_vpc_20100810/en/index.html.
Chan, M. and Ki-Moon, B. (24/09/09), *Support for Developing Countries' Response to the H1N1 Influenza Pandemic*, Joint Statement by the UN Secretary-General and the WHO Director-General, available at: http://www.who.int/mediacentre/news/statements/2009/h1n1_support_20090924/en/index.html.
FOA/WHO/OIE (07/05/09), *Joint FOA/WHO/OIE Statement on Influenza A(H1N1) and the Safety of Pork*, Joint FOA/WHO/OIE Statement, available at: http://www.who.int/mediacentre/news/statements/2009/h1n1_20090430/en/index.html.
WHO/IRRC/UNSIC/OCHA/UNICEF (17/08/09), *Call to Action from WHO, IFRC, UNISIC, OCHA, and UNICEF*, available at: http://www.who.int/csr/resources/publications/swineflu/call_action/en/index.html.

WHO Briefing Notes

WHO (08/07/09), *Viruses Resistant to Osetamivir (Tamiflu) Identified*. Pandemic (H1N1) Briefing Note 1, available at: ttp://www.who.int/csr/disease/swineflu/notes/en/index.html.
WHO (13/07/09), *WHO Recommendations on Pandemic (H1N1) 2009 Vaccine*. Pandemic (H1N1) Briefing Note 2, available at: http://www.who.int/csr/disease/swineflu/notes/en/index.html.
WHO (16/07/09), *Changes on Reporting Requirements for Pandemic (H1N!) 2009 Virus Infection*. Pandemic (H1N1) Briefing Note 3, available at: http://www.who.int/csr/disease/swineflu/notes/en/index.html.
WHO (24/07/09), *Preliminary Information Important for Understanding the Evolving Situation*. Pandemic (H1N1) Briefing Note 4, available at: http://www.who.int/csr/disease/swineflu/notes/en/index.html.
WHO (31/07/09), *Pandemic Influenza in Pregnant Women*. Pandemic (H1N1) Briefing Note 5, available at: http://www.who.int/csr/disease/swineflu/notes/en/index.html.
WHO (06/08/09), *Safety of Pandemic Vaccines*. Pandemic (H1N1) Briefing Note 6, available at: http://www.who.int/csr/disease/swineflu/notes/en/index.html.
WHO (06/08/07b), *Pandemic Influenza Vaccines Manufacturing Process and Timeline*. Pandemic (H1N1) Briefing Note 7, available at: http://www.who.int/csr/disease/swineflu/notes/en/index.html.
WHO (20/08/09), *Recommended Use of Antivirals*. Pandemic (H1N1) Briefing Note 8, available at: http://www.who.int/csr/disease/swineflu/notes/en/index.html.
WHO (28/08/09), *Preparing for the Second Wave: Lessons from Current Outbreak*. Pandemic (H1N1) Briefing Note 9, available at: http://www.who.int/csr/disease/swineflu/notes/en/index.html.
WHO (11/09/09), *Measures in School Settings*. Pandemic (H1N1) Briefing Note 10, available at: http://www.who.int/csr/disease/swineflu/notes/en/index.html.

WHO (16/10/09), *Clinical Features of Severe Cases of Pandemic Influenza*. Pandemic (H1N1) Briefing Note 13, available at: http://www.who.int/csr/disease/ swineflu/notes/en/index.html.
WHO (24/10/09), *Pandemic Influenza Vaccines: Current Status*. Pandemic (H1N1) Briefing Note 11, available at: http://www.who.int/csr/disease/swineflu/notes/ en/index.html.
WHO (25/10/09), *Antiviral Use and the Risk of Drug Resistance*. Pandemic (H1N1) Briefing Note 12, available at: http://www.who.int/csr/disease/swineflu/notes/ en/index.html.
WHO (30/10/09), *Experts Advise WHO on Pandemic Vaccine Policies and Strategies*. Pandemic (H1N1) Briefing Note 14, available at: http://www.who.int/csr/ disease/swineflu/notes/en/index.html.
WHO (05/11/09), *Infection of Farmed Animals with the pandemic Virus*. Pandemic (H1N1) Briefing Note 15, available at: http://www.who.int/csr/disease/ swineflu/notes/en/index.html.
WHO (19/11/09), *Safety of Pandemic Vaccines*. Pandemic (H1N1) Briefing Note 16, available at: http://www.who.int/csr/disease/swineflu/notes/en/index.html.
WHO (20/11/09), *Public Health Significance of Virus Mutation Detected in Norway*. Pandemic (H1N1) Briefing Note 17, available at: http://www.who.int/csr/ disease/swineflu/notes/en/index.html.
WHO (02/12/09), *Oseltamivir Resistance in Immunocompromised Hospital Patients*. Pandemic (H1N1) Briefing Note 18, available at: http://www.who.int/csr/ disease/swineflu/notes/en/index.html.
WHO (03/12/09), *WHO Use of Advisory Bodies in Responding to the Influenza Pandemic*. Pandemic (H1N1) Briefing Note 19, available at: http://www.who.int/ csr/disease/swineflu/notes/en/index.html.
WHO (22/12/09), *Comparing Deaths from Pandemic and Seasonal influenza*. Pandemic (H1N1) Briefing Note 20, available at: http://www.who.int/csr/disease/ swineflu/notes/en/index.html.
WHO (10/06/10), *The International Response to the Influenza Pandemic: WHO Responds to the Critics*. Pandemic (H1N1) Briefing Note 21, available at: http:// www.who.int/csr/disease/swineflu/notes/en/index.html.
WHO (21/07/10), *Monitoring Patterns and Levels of Worldwide Activity*. Pandemic (H1N1) Briefing Note 22, available at: http://www.who.int/csr/disease/ swineflu/notes/en/index.html.
WHO (10/08/10), *WHO Recommendations for the Post-Pandemic Period*. Pandemic (H1N1) Briefing Note 23, available at: http://www.who.int/csr/disease/ swineflu/notes/en/index.html.

WHO Press Briefings

Ben Embarek, P. (03/05/09), *WHO Press Briefing 03/05/09*, available at: http:// www.who.int/mediacentre/multimedia/swineflupressbriefings/en/index .html.
Ben Embarek, P. (04/05/09), *WHO Press Briefing 04/05/09*, available at: http:// www.who.int/mediacentre/multimedia/swineflupressbriefings/en/index.html.
Briand, S. (08/05/09), *WHO Press Briefing 08/05/09*, available at: http://www.who .int/mediacentre/multimedia/swineflupressbriefings/en/index.html.

232   Bibliography

Briand, S. (13/05/09), *WHO Press Briefing 13/05/09*, available at: http://www.who.int/mediacentre/multimedia/swineflupressbriefings/en/index.html.
Chan, M. (29/04/09), *WHO Press Briefing 29/04/09b*, available at: http://www.who.int/mediacentre/multimedia/swineflupressbriefings/en/index.html.
Chan, M. (11/06/09b), *WHO Press Briefing 11/06/09*, available at: http://www.who.int/mediacentre/multimedia/swineflupressbriefings/en/index.html.
Chan, M. (10/08/10), *WHO Press Briefing 10/08/10*, available at: http://www.who.int/mediacentre/multimedia/swineflupressbriefings/en/index.html.
Fineberg, H. (19/05/10), *WHO Press Briefing 19/05/10*, available at: http://www.who.int/mediacentre/multimedia/swineflupressbriefings/en/index.html.
Fineberg, H. (02/07/10), *WHO Press Briefing 02/07/10*, available at: http://www.who.int/mediacentre/multimedia/swineflupressbriefings/en/index.html.
Fukuda, K. (26/04/09), *WHO Press Briefing 26/04/09*, available at: http://www.who.int/mediacentre/multimedia/swineflupressbriefings/en/index.html.
Fukuda, K. (27/04/09), *WHO Press Briefing 27/04/09*, available at: http://www.who.int/mediacentre/multimedia/swineflupressbriefings/en/index.html.
Fukuda, K. (27/04/09b), *WHO Press Briefing 27/04/09b*, available at: http://www.who.int/mediacentre/multimedia/swineflupressbriefings/en/index.html.
Fukuda, K. (28/04/09), *WHO Press Briefing 28/04/09*, available at: http://www.who.int/mediacentre/multimedia/swineflupressbriefings/en/index.html.
Fukuda, K. (29/04/09), *WHO Press Briefing 29/04/09b*, available at: http://www.who.int/mediacentre/multimedia/swineflupressbriefings/en/index.html.
Fukuda, K. (30/04/09), *WHO Press Briefing 30/04/09*, available at:http://www.who.int/mediacentre/multimedia/swineflupressbriefings/en/index.html.
Fukuda, K. (01/05/09), *WHO Press Briefing 01/05/09*, available at: http://www.who.int/mediacentre/multimedia/swineflupressbriefings/en/index.html.
Fukuda, K. (03/05/09), *WHO Press Briefing 03/05/09*, available at: http://www.who.int/mediacentre/multimedia/swineflupressbriefings/en/index.html.
Fukuda, K. (04/05/09), *WHO Press Briefing 04/05/09*, available at: http://www.who.int/mediacentre/multimedia/swineflupressbriefings/en/index.html.
Fukuda, K. (05/05/09), *WHO Press Briefing 05/05/09*, available at:http://www.who.int/mediacentre/multimedia/swineflupressbriefings/en/index.html.
Fukuda, K. (06/05/09), *WHO Press Briefing 06/05/09*, available at: http://www.who.int/mediacentre/multimedia/swineflupressbriefings/en/index.html.
Fukuda, K. (07/05/09), *WHO Press Briefing 07/05/09*, available at: http://www.who.int/mediacentre/multimedia/swineflupressbriefings/en/index.html.
Fukuda, K. (11/05/09), *WHO Press Briefing 11/05/09*, available at: http://www.who.int/mediacentre/multimedia/swineflupressbriefings/en/index.html.
Fukuda, K. (14/05/09), *WHO Press Briefing 14/05/09*, available at: http://www.who.int/mediacentre/multimedia/swineflupressbriefings/en/index.html.
Fukuda, K. (14/05/09b), *WHO Press Briefing 14/05/09*, available at: http://www.who.int/mediacentre/multimedia/swineflupressbriefings/en/index.html.
Fukuda, K. (26/05/09), *WHO Press Briefing 26/05/09*, available at: http://www.who.int/mediacentre/multimedia/swineflupressbriefings/en/index.html.
Fukuda, K. (02/06/09), *WHO Press Briefing 02/06/09*, available at: http://www.who.int/mediacentre/multimedia/swineflupressbriefings/en/index.html.
Fukuda, K. (07/06/09), *WHO Press Briefing 07/06/09*, available at: http://www.who.int/mediacentre/multimedia/swineflupressbriefings/en/index.html.

Fukuda, K. (09/06/09), *WHO Press Briefing 09/06/09*, available at: http://www.who.int/mediacentre/multimedia/swineflupressbriefings/en/index.html.
Fukuda, K. (07/07/09), *WHO Press Briefing 07/07/09*, available at: http://www.who.int/mediacentre/multimedia/swineflupressbriefings/en/index.html.
Fukuda, K. (24/09/09), *WHO Press Briefing 24/09/09*, available at: http://www.who.int/mediacentre/multimedia/swineflupressbriefings/en/index.html.
Fukuda, K. (05/11/09), *WHO Press Briefing 05/11/09*, available at: http://www.who.int/mediacentre/multimedia/swineflupressbriefings/en/index.html.
Fukuda, K. (26/11/09), *WHO Press Briefing 26/11/09*, available at: http://www.who.int/mediacentre/multimedia/swineflupressbriefings/en/index.html.
Fukuda, K. (03/12/09), *WHO Press Briefing 03/12/09*, available at: http://www.who.int/mediacentre/multimedia/swineflupressbriefings/en/index.html.
Fukuda, K. (17/12/09), *WHO Press Briefing 17/12/09*, available at: http://www.who.int/mediacentre/multimedia/swineflupressbriefings/en/index.html.
Fukuda, K. (14/01/10), *WHO Press Briefing 14/01/10*, available at: http://www.who.int/mediacentre/multimedia/swineflupressbriefings/en/index.html.
Fukuda, K. (11/02/10), *WHO Press Briefing 11/02/10*, available at: http://www.who.int/mediacentre/multimedia/swineflupressbriefings/en/index.html.
Fukuda, K. (14/02/10), *WHO Press Briefing 14/02/10*, available at:http://www.who.int/mediacentre/multimedia/swineflupressbriefings/en/index.html.
Fukuda, K. (18/02/10), *WHO Press Briefing 18/02/10*, available at: http://www.who.int/mediacentre/multimedia/swineflupressbriefings/en/index.html.
Fukuda, K. (24/02/10), *WHO Press Briefing 24/02/10*, available at: http://www.who.int/mediacentre/multimedia/swineflupressbriefings/en/index.html.
Fukuda, K. (20/03/10), *WHO Press Briefing 29/03/10*, available at: http://www.who.int/mediacentre/multimedia/swineflupressbriefings/en/index.html.
Härtl, G. (27/04/09), *WHO Press Briefing 27/04/09b*, available at: http://www.who.int/mediacentre/multimedia/swineflupressbriefings/en/index.html.
Kieny, M-P. (01/05/09), *WHO Press Briefing 01/05/09*, available at: http://www.who.int/mediacentre/multimedia/swineflupressbriefings/en/index.html.
Kieny, M-P. (06/05/09), *WHO Press Briefing 06/05/09*, available at: http://www.who.int/mediacentre/multimedia/swineflupressbriefings/en/index.html.
Kieny, M-P. (13/07/09), *WHO Press Briefing 13/07/09*, available at: http://www.who.int/mediacentre/multimedia/swineflupressbriefings/en/index.html.
Kieny, M-P. (06/08/09), *WHO Press Briefing 06/07/09*, available at: http://www.who.int/mediacentre/multimedia/swineflupressbriefings/en/index.html.
Kieny, M-P. (24/09/09), *WHO Press Briefing 24/09/09*, available at: http://www.who.int/mediacentre/multimedia/swineflupressbriefings/en/index.html.
Kieny, M-P. (30/10/09), *WHO Press Briefing 30/10/09*, available at: http://www.who.int/mediacentre/multimedia/swineflupressbriefings/en/index.html.
Kieny, M-P. (19/11/09), *WHO Press Briefing 19/11/09*, available at: http://www.who.int/mediacentre/multimedia/swineflupressbriefings/en/index.html.
Ryan, M. (02/05/09), *WHO Press Briefing 02/05/09*, available at: http://www.who.int/mediacentre/multimedia/swineflupressbriefings/en/index.html.
Shindo, N. (12/05/09), *WHO Press Briefing 12/05/09*, available at: http://www.who.int/mediacentre/multimedia/swineflupressbriefings/en/index.html.
Shindo, N. (12/11/09), *WHO Press Briefing 12/11/09*, available at: http://www.who.int/mediacentre/multimedia/swineflupressbriefings/en/index.html.

WHO Pandemic Preparedness Document

WHO (2009), *Pandemic Influenza Preparedness and Response: A WHO Document*. Geneva: Global Influenza Programme, Health Security and Environment Cluster, WHO.

Council of Europe Documents

Council of Europe Parliamentary Assembly (24/06/10), *Verbatim Report – Twenty-Sixth Sitting of the Parliamentary Assembly of the Council of Europe*, available at: http://assembly.coe.int/Main.asp?/Documents/Records/2010/E/10062441500.htm.

Council of Europe PACE Meeting (29/03/10), *Questions and Debate*, available at: http://www.assembly.coe.int/CommitteeDocs/2010/20122329.

Flynn, P. (23/03/10), *Memorandum: The Handling of the H1N1 Pandemic: More Transparency Needed*, available at: http://www.assembly.coe.int/CommitteeDocs/2010/20122329_MemorandumPandemic_E.htm.

Flynn, P. (29/03/10), *Speech – Paul Flynn, Rapporteur*, available at: mms://coenews.int.vod/100329_w01_w.wnv.

Flynn, P. (07/06/10), *The Handling of the H1N1 Pandemic: More Transparency Needed*. [Doc no. 12283 – Passed by the Council of Europe 24/06/10] Strasbourg: Social Health and Family Affairs Committee, Council of Europe.

Fukuda, K. (26/01/10), *Statement by Dr Keiji Fukuda on Behalf of WHO at the Council of Europe Hearing on Pandemic (H1N1) 2009*, available at: http://www.coe.int/t/DC/Files/PA_session/jan_2010.

Gentilini, M. (29/03/10), *Speech – Marc Gentilini*, available at: mms://coenews.int.vod/100329_w01_w.wnv.

Hessel, L. (26/01/10), *EVM Statement to the Council of Europe Hearing 'The Handling of Pandemic Preparedness: More Transparency Needed?' On the Motion 'Faked Pandemic – A Threat to Health'*, available at: http://www.coe.int/t/DC/Files/PA_session/jan_2010.

Jefferson, T. (29/03/10), *Speech – Dr. Tom Jefferson, The Cochrane Collaboration*, available at:mms://coenews.int.vod/100329_w01_w.wnv.

Keil, U. (26/010/10), *Introductory Statement by Prof. Dr. Ulrich Keil*, available at: http://www.coe.int/t/DC/Files/PA_session/jan_2010.

Kopacz, E. (29/03/10), *Address of Ms. Ewa Kopacz, Minister of Health of Poland*, available at: mms://coenews.int.vod/100329_w01_w.wnv.

Rivasi, M. (29/03/10), *Speech – Michele Rivasi*, available at: mms://coenews.int.vod/100329_w01_w.wnv.

Wodarg, W. (18/12/09), *Faked Pandemics – A threat to Health*, Motion of a Recommendation by the Parliamentary Assembly of the Council of Europe Doc. 122110. Strasbourg: Council of Europe.

Wodarg, W. (26/01/10), *Hearing on 'The Handling of the H1N1 Pandemic' More Transparency Needed?*, available at: http://www.coe.int/t/DC/Files/PA_session/jan_2010.

Index

accountability, 6, 134, 162, 174, 194
actor networks, 6–8, 13, 16, 20–2, 25, 29, 31–2, 52, 103, 133, 135, 137, 163, 171, 205, 207–8
actor-network theory (ANT), 6–8, 32
adverse reactions of vaccines, 115, 118, 120–3, 134, 149, 161–3
air travel, 127, 173, 178, 200
  see also travel restrictions
animal viruses, 24–5, 28, 213
anti-flu vaccines, *see* vaccines/vaccination
antivirals
  access and availability, 124
  dismissal of, 124–5
  over-the-counter, 194
  profit-making interventions, 103
  resistance to, 124–5, 194–5
  as a secondary measure, 123, 131
  stock, 78
  studies, 160
  utility and efficacy, 102, 123–5
anxiety, 25, 28, 95, 146–7
Asian influenza, *see* H2N2
authority, regulatory, 5, 7, 51, 60, 80, 107, 117, 136, 146, 160, 163–4, 174–5, 180–1, 184, 189, 194–6
availability of vaccines, 106, 111–14, 157, 193, 199–200
avian influenza, *see* H5N1

bird flu, *see* H5N1
black-boxed concepts, 7–13, 35–7, 39, 41–2, 44, 68, 74, 173, 204–5, 207–8, 214
blame, 25, 29, 125, 147, 150, 154–5, 157, 159, 169–70, 195, 197
border control, 2, 102, 123
  and quarantine, 125–30
boundaries of authority, 136, 174, 196
bureaucracy, 103, 105–6, 150

categorization of H1N1, 64–101, 206
civil society, 180, 187
classification of H1N1, 64–101
classificatory schemes, 66, 68–71, 76, 79–81, 84, 89, 101, 202
climate change, 52, 174–5, 209
clinical severity, 35–45, 50, 97–8, 159, 208
  see also severity
clinical trials, 116–18, 123, 160–1, 163
closure of borders and quarantine, 125–30
collaboration, 24, 70–1, 97, 111, 183, 185–9
collective memory, 16, 205
collective responses to infectious disease, 125
common cold, 36, 124
community-level transmission, 77–80, 82, 94, 210, 213
complications
  of H1N1 vaccine, 120, 161–2
  of H1N1 virus, 39, 42
conflicts of interest, 136, 164–5
constructionist approaches, 32, 34, 66–7, 144, 208
containment, 76, 103, 125–6, 149, 210, 213
contemporary science, 33, 133, 167, 202
contested expertise, 163–8
co-productionist theory, 31–5, 47, 50–1, 71, 103, 122, 136, 174–5, 202
corporate responsibility, 111
Council of Europe's critique of WHO's policy, 2–5, 52, 103, 115, 120, 131–2, 133–71, 172, 174, 191, 194–5, 204, 206–8, 211–13
'crying wolf,' 149–51
definition of 'pandemic,' 140–4

Council of Europe's critique of WHO's policy – *continued*
  'Faked Pandemics: A Threat to Public Health,' 134
  management of H1N1, 168–71
  nature of influenza and H1N1, 137–40
  pandemic phase declarations and definitions, 74–83, 152–5
  risk narrative, 144–8
  risk and trust, 148–52
  scientific expertise, 163–8
  transparency, 151–2
  vaccine use, 156–63
  Wodarg's assertions, 134–5
Council of Europe's representatives, 212–13
  see also specific types
credibility, 68, 103, 109, 131, 134, 157, 161, 163
'crying wolf,' critique of, 149–51

data collection, 46–50, 59, 85, 100, 105, 118
deaths
  avian flu, 74
  H1N1, 2, 39, 41, 56, 58, 60–1, 141, 160, 195, 210–12
  Hong Kong flu, 21
  influenza-like illness/flu, 138
  pandemic influenza, 12, 21, 58
  SARS, 74
  seasonal flu, 14–15, 58, 141
  Spanish flu, 21
  vaccine-induced, 115, 122, 161
decision-making, 5, 47, 62–3, 80, 103–5, 108, 115, 122, 144, 147, 161–2, 166, 172–3, 180, 182, 184, 193–4
declaration of H1N1 pandemic, 1–2, 10, 12–13, 28–9, 45, 63, 65, 71, 74–83, 205
  criticism of, 142–4
democratization of science, 4, 133, 136, 163

Director-General of WHO, 10–11, 46, 111, 113, 125–6, 185, 187, 197, 211–12
  Chan, Margaret, 9–11, 18, 21–2, 27, 39, 43, 46–7, 50, 83, 85–6, 88–90, 110, 130, 174, 178, 185–8, 192, 197–8, 211–12
discursive path dependency, 104–5
disease control measures, 128–30, 200
division of authority, 174, 196

economic impact, 1, 97, 131, 148, 156, 162, 173, 176, 178–9, 186
Emergency Committee, 79, 129, 210–11
enrolment (ANT), 7, 29
epidemics, 19–20, 28, 56, 72–3, 86, 126, 170, 176, 179, 183
  see also specific types
epidemiology, 11–12, 14, 18–20, 34–5, 37–40, 46–9, 51, 53–4, 56–7, 60–2, 65, 72, 77, 98, 131, 134, 140–1, 149, 164, 193, 210
epistemic communities, 52
expertise, WHO's contested, 163–8
  advice for policy-making, 166, 168
  code of conduct, 166
  decision-making, 166
  favouring pharmaceutical manufacturers, 166–7
  as key opinion leaders (KOLs), 167
  lack of transparency, 164–6
  as manufactured entities, 167
  and nature of risks, 168
  privacy issues, 165
experts, 29, 35, 68, 70–1, 91–3, 97, 136–7, 143, 152–3, 159–60, 170–1, 174, 185, 190
external actors, 3–5, 7, 10, 23, 29, 66, 114–15, 133, 172–3, 202, 204, 206

fact-making, 6, 57, 103, 156, 165, 209
fatality, 18, 42, 58–9
fear, 19, 28, 95, 145, 147, 158, 179, 200
Flynn, Paul, 135, 139, 141–56, 158–60, 165, 169, 212
funding, 33, 156, 159, 166, 180

Index   237

genetic mutation, 8, 10, 53–5, 150, 188
geographical spread, 9, 11–12, 12, 16, 20, 32, 35, 72–3, 75–7, 77, 79, 92–5, 94, 97, 100–1, 173, 206
global disease threats, 1, 6, 21, 175, 177–9, 208–9
global governance, 112, 176, 191
see also governance
Global Influenza Programme, 65, 137–8
globalization and global public health, 32, 172–203
 coordination, cooperation and collaboration, 179–90, 207–8
 definition of globalization, 173
 developing vs. developed countries, vulnerability of, 196–200
 and global health paradigm, 175–7, 207–8
 H1N1 as a global threat, 177–9
 international hostilities caused by H1N1, 200–1
 multiple actors and stakeholders, 179–90, 207–8
 severity and risk assessment, 192–3
 and transparency in global public health, 202–3
 vulnerability and impact of infectious diseases, 173–4
 WHO and national governments, 190–6, 208
 WHO's new role and actions, 179–90, 207–8
global outbreak, 9, 49–50, 73, 76, 88–91
global policy, 89–91, 99–100, 176, 191–2
global politics, 174
global public health, 1, 4–5, 49, 71, 102, 106, 110, 114, 131–2, 134, 136, 152, 204, 207–8
see also globalization and global public health
global public-private partnerships (GPPPs), 181, 186–7
global risk management, 112
global scale, 14, 176, 199
global solidarity, 187–8

global surveillance, 181
governance, 14, 112, 131, 136–7, 176, 179–80, 191, 208

H1N1 pandemic (swine flu; 2009)
 analogy to past pandemics, 16–22
 comparisons with previous pandemics, 145–6, 205
 contemporary threats, 21–2
 declaration, 1–2, 10, 12–13, 28–9, 45, 63, 65, 71, 74–83, 205
 vs. H5N1 avian influenza, 21–2, 26, 146, 148–9
 as a global threat, 177–9
 mortality rates, 2, 39, 41, 56, 58, 60–1, 141, 160, 195, 210–12
 origins and zoonosis, 22–7
 vs. seasonal influenza, 13–16, 39–40, 85–6, 88, 141, 205
 severity, 20–1, 35–45, 145–50
 timeline of events/summary, 210–13
 see also management of H1N1; nature of H1N1; pandemic alert phases; risk of H1N1
H1N1 subtype, 143, 149, 152
H2N2 Asian influenza (1957), 18
H3N2 Hong Kong Influenza (1968/1969), 2, 18, 21, 85
H5N1 avian influenza (2004–2006), 1, 17, 21–2, 26, 36, 73–4, 113, 124, 146, 148, 150–1, 156–7, 169, 179
health systems, 78, 131, 176, 181, 186
high-risk groups, 37, 98, 124, 146
historical analogy, 16–22, 83–4, 149–50, 175, 198, 205
historical path dependency, 104–5
HIV/AIDS, 17, 36–7, 150
Hong Kong influenza, see H3N2
hospitals/hospitalization, 42, 62, 98
How Institutions Think (Douglas), 105–6
human rights, dignity and fundamental freedoms, 201
human-to-human transmission, 76, 210, 213

IHR (International Health Regulations), 129, 181, 201, 213
immunity, 9, 15, 86, 99, 139–41

Index

immunization, *see* vaccines/vaccination
immunological complications, 9, 162
infectious disease
  antivirals, 194–5
  border control and quarantine, 125, 128
  governance, 131
  historical analogy, 16–21, 146, 205, 207
  impact of globalization, 173–7
  institutionalized reaction to, 4, 10, 184, 208
  management programmes, 106–9
  origin of, 22–5, 196–8
  risk/severity, 36, 43–4, 209
  threats, 1–3, 34, 47, 73, 109, 130, 177, 179, 188, 190–1
  vaccines, 103, 105–9, 115, 119, 122, 131, 149
  WHO's historical successes against, 103, 105–9, 131
influenza
  antivirals, *see* antivirals
  Asian influenza (H2N2; 1957), 18
  avian, 1, 17, 21–2, 26, 36, 73–4, 113, 124, 146, 148–51, 156–7, 169, 179
  genetic mutation, 8, 10, 53–5, 150, 188
  global outbreak, 1, 9–11
  Hong Kong, 2, 18, 21, 85
  vs. ILIs/flu, 137–9
  pandemic, 1–2, 9, 11–14, 17–19, 37, 39, 44, 58–9, 64, 68, 71, 83, 89, 101, 107–8, 115, 118, 130–1, 139, 145, 155, 164, 166, 171, 178–9, 185, 187, 193, 197, 205, 210
  resistance to heat, 25
  seasonal, 6, 8, 12–16, 29, 37–40, 45, 58–9, 78, 84–6, 88, 90–1, 107, 110, 114, 116, 118–19, 121, 140–1, 155, 162, 166, 205, 210, 213
  Spanish, 2, 17–21, 58–9, 144–6, 149, 175
  surveillance programmes, 137
  vaccines, *see* vaccines/vaccination

influenza-like illnesses/flu (ILIs), 16, 145–5
  vs. influenza, 137–9
  severe respiratory symptoms and, 140
information gathering, 35, 42, 46–57, 59, 62, 65, 79, 92, 99–100, 105, 128–9, 151, 164–5, 168, 170, 172–3, 181–2, 184–5, 189, 191–2, 194–5, 201–3, 207–8, 210
inoculation, *see* vaccines/vaccination
institutional path dependency, 4, 105–6, 115
interessement (ANT), 7, 29
international governance, 176
  see also global governance
international health *vs*. global health, 24, 175–7
international hostilities caused by H1N1, 200–1
isolation, 125, 128, 177, 196

labelling, 27, 87, 140–1, 153, 205
laboratories, 2, 58, 60–1, 121, 158, 210–11
local actions, 73, 76, 91, 193
local outbreaks, 73, 88–9, 91, 175, 192–3

management of H1N1 (WHO's), 102–32, 207
  antivirals, 123–5
  border control and quarantine, 125–30
  criticism of, 102–31, 156–71
  institutional processes, 101–6; institutional origin/change, 104–6; new institutionalism, 103–4; organizational structures, 103, 106; path dependency, concept of, 104–6
  monitoring/surveillance, 130–2
  other health measures, 108, 123–32
  preventative strategies, 102–32
  vaccines, 102–3, 106–23, 156–63
mandates, 108, 114, 117, 180
mass vaccination, 5, 103, 106–9, 115, 117, 119–20, 122, 131, 139, 159, 175, 183, 199, 207

media, 23, 38, 54, 58, 60, 83, 88, 121–2, 145, 148, 163, 184, 200, 202, 212
medicine supplies, 162, 165, 199
Mexico
  H1N1 outbreak, 46, 76–7, 142–3, 210–11
  international hostilities against, 200
mildness, 2, 4, 13–16, 18, 20–2, 35–45, 48, 50, 55–6, 58–9, 63, 72, 74, 82, 84, 86, 92, 96, 124, 133, 140–2, 145, 149, 151, 153, 157–9, 168, 198
mixed virus pattern, 85–6, 95
mobilization, 6–8, 13–14, 17, 19, 29–30, 31–2, 38, 44–5, 47, 51–2, 63, 78, 104, 122, 144, 168, 183, 189–90, 205
monitoring and surveillance, 1, 9, 12, 14, 47, 53–4, 61, 71, 77, 90, 116, 123, 130–2, 184, 189, 193
morbidity and mortality rates, 2, 54–5, 60, 74, 141, 206
mortality rates, 2, 14, 44, 54, 60–2, 74, 93, 139, 141, 206
mutation, genetic, 8, 10, 53–5, 150, 188

naming, 27–30
national government responses *vs.* global health paradigm, 190–6
  reporting, 195
  severity and risk assessment, 192–3
  WHO's response to, 194–5
national governments, 7, 25, 61, 69, 71, 100, 102, 112, 123, 128, 145, 154, 161–2, 174, 184, 202–3, 208–9
National Pandemic Preparedness Plan, 2, 123, 157
nation states, 174, 176, 180, 187, 190–1, 194, 202
nature of H1N1 (WHO's construction), 13–30
  Council of Europe's account, 137–40
  criticism of, 6–30, 168–71
  definition of 'pandemic,' 1–3, 7–13, 204–5; criticism of, 140–4
  historical analogy, 16–22
  naming, 27–30
  origins and zoonosis, 22–7; role of pigs, 24–7
  vs. seasonal influenza, 13–16, 37
  new disease, 10, 16, 22–3, 27–8, 46–8, 129
new institutionalism, 103–4, 180
new structuring of the WHO, 179–86
non-governmental organizations (NGOs), 102, 112, 176, 180, 187
non-pandemic disease, 7, 12
non-pandemic period, 83–4, 91
non-state actors, 176, 180, 187, 191
novelty of viral strains, 5, 8–10, 12–13, 15–16, 20, 24, 28, 40, 45–6, 57, 66, 74, 82, 93, 116, 139–40

origins and zoonosis of H1N1, 22–7, 196
oseltamivir, 124, 194

PACE (Parliamentary Assembly of the Council of Europe), 150, 157, 211, 213
pandemic alert phases, 13, 64–101, 206–7
  criticisms of, 64–101, 152–5, 206–7
  declaration and definition, 74–83
  function of, 64–74
  pharmaceutical corporations' influence/role, 156–63
  Phase 3, 76, 79, 210
  Phase 4, 76–7, 125–6, 210
  Phase 5, 77–8, 93–4, 97, 210–11
  Phase 6, 65, 68–70, 72, 74–5, 77–9, 81–3, 86, 88–90, 93–5, 97, 100, 143, 152–4, 157, 159–60, 211–12
  vs. post-pandemic declaration and uncertainty, 83–92
  redefinitions, inclusion of severity, 92–101, 152–5
  summary, 210–13
  timeline of events, 210–12
Pandemic Influenza Preparedness and Response (2009), 65
Pandemic Planning Guidance, 96
Pandemic Preparedness Guidelines (2009), 70, 91

pandemic preparedness plans, 2, 123, 157, 211
pandemic, WHO characterizations, 7–13
  Council of Europe' narrative, 137–44
  criticism of, 137–44
  geographical spread, 9–10
  long-term monitoring and action, 12
  pandemic *vs.* seasonal influenza, 12, 37, 39, 205
  spectrum of severity, 43–4
  susceptibility of global populations, 11–12
  unpredictability/uncertainty, 10–11
  viral novelty, 8–9
  see also nature of H1N1
panic, 80, 82, 129, 135, 148–52
partnerships and cooperation, WHO, 179–90
past pandemics, 16–22
  Asian influenza (1957), 18
  contemporary threats, 21–2
  estimate of severity, 20–1
  globalized nature of, 177–9
  H3N2 Hong Kong Influenza (1968/1969), 2, 18, 21, 85
  H5N1 avian influenza (2004–2006), 1, 17, 21–2, 26, 36, 73–4, 113, 124, 146, 148, 150–1, 156–7, 169, 179
  HIV/AIDS, 17, 36–7
  impact on global economies, 179
  influenza pandemics, 17
  SARS (2003), 17, 179
  Spanish influenza (1918), 2, 17–21, 58–9, 144–6, 149, 175
path dependency, 4–5, 104–6, 109, 115, 122–3, 131–2
pharmaceutical corporations, 78, 102, 109–15
  access/distribution of vaccines, 2, 106, 110–14, 118
  capacity for vaccine production, 109–10, 114–15; and timing of production, 110–11
  corporate responsibility, 111

cost/payment, 113, 159–60
criticism of, 111–15, 134–5, 148, 153, 156–63
expertise and know-how, 190
influence/role, 111–15, 135, 142, 156–63, 167–9
interrelations with multiple actors, 111–15, 166–7
novel methods of production, 162–3
profit-motivated influence, 102, 113–14, 156–63
regulatory measures, 118
vested interest of, 157–8
WHO's collusion with, 111–15, 156–63
and WHO's new paradigm, 181–3, 189–90
Phase 3, 76, 79, 210
Phase 4, 76–7, 125–6, 210
Phase 5, 77–8, 93–4, 97, 210–11
Phase 6, 65, 68–70, 72, 74–5, 77–9, 81–3, 86, 88–90, 93–5, 97, 100, 143, 152–4, 157, 159–60, 211–12
pneumonia, 39, 86, 161
policy-making/policy-makers, 51–2, 103, 136–7, 168, 186, 191
political contexts, 4, 46, 52, 68, 98, 102, 104, 108, 111, 131, 133, 135–6, 140, 154, 163, 165–6, 168, 173–4, 176, 180, 186, 208
political economy approaches, 102, 113
post-pandemic period, 74, 83–6, 88, 90, 92, 212
post-peak period, 84, 89, 91
precautionary principle, 52, 146–8
pre-emptive health measures, 1, 75, 111, 125, 168
pre-existing conditions, 9, 16, 99–100, 113, 175, 204, 206
premature declaration/predeclaration, 74, 82, 142, 152–5
preparedness, pandemic, 2, 17, 19, 21, 41, 49, 56, 65, 70, 73, 91, 100, 123, 128, 157, 169, 181–2, 184, 197, 211
preventative strategies, 102–32
  institutional processes, 102–6
  other measures, 123–32

vaccines, 102–3, 106–23
   see also management of H1N1
      (WHO's)
private-public partnerships, 176,
   180
private sector, 112, 176–7, 181–3,
   186–7, 189–90
problematization (ANT), 7–8,
   13–14, 16–22, 25, 27, 29, 101,
   139, 205
public health measures, alternative, 1,
   108, 123, 131, 161, 201
public health priorities, 150, 196
public-private partnerships, 177,
   181, 186
punctualization (ANT), 6–7, 11

quality of vaccines, 121, 160
quarantine and border control, 123,
   125–30, 196
   see also travel restrictions

rapid containment, 126, 213
recommendations, 89, 112, 129, 134,
   147, 156, 160–1, 165, 172, 181,
   192–3, 200, 211
recovery, 39, 42–3, 121, 145–6
redefinitions, 36–8, 40, 42, 45, 65,
   68–9, 92–101
reflexive governance of knowledge,
   32, 136–7
regulatory authority, 5, 7, 51, 60, 80,
   107, 117, 136, 146, 160, 163–4,
   174–5, 180–1, 184, 189,
   194–6
responsibility, 61, 71–2, 92, 111, 145,
   147, 150–1, 159, 162, 169, 172,
   184–6, 193, 195, 197, 202
restrictions of travel, 125–7, 129
risk management, 24, 51, 103–4, 112,
   116, 133–4, 146–7, 185
risk of H1N1 (WHO's construction),
   31–63
   co-productionist explanations,
      32–5
   Council of Europe's account, 144–52
   criticism of, 31–63, 144–52
   evolving threat and scientific
      uncertainty, 51–7, 131, 205–6
   and severity/mildness, 35–45,
      145–50
   sociological accounts of risk, 32–5,
      148
   statistical uncertainty, 57–63
   and trust, 148–52
   uncertainty and lack of
      information, 45–51
risk society, 32–4, 47

safety
   exposure to pigs, 25–6
   of travellers, 126–7
   vaccine, 102, 107–8, 115–20, 122,
      124, 131, 160
SARS (severe acute respiratory
   syndrome; 2003), 1, 17, 19–20,
   73–4, 146, 148–51, 179, 183
science–policy interface, 50, 52,
   57, 136
science, structuring of, 32–4, 163,
   174
scientific expertise, 70–1, 97, 137,
   153, 163–8
scientific fact-making, 103
scientific facts, 1, 3, 6, 30, 46, 48, 57,
   75, 103, 133, 136–7, 139–40, 154,
   156, 165, 168, 171, 202, 204–5,
   207–8
scientific institutions, 32, 136–7,
   206
scientific justification, 147
scientific knowledge, 5, 31–4, 46,
   52–3, 67, 101, 108, 166
scientific research, 4, 33–4, 202
seasonal influenza, 8
   and H1N1 pandemic, comparison,
      13–16, 39, 85–6, 88, 141
   and H1N1 vaccines, 116, 119
   mortality rates, 14–15, 141
   vs. pandemic influenza, 12, 38, 86,
      155, 160
   post-pandemic period, 85–6
   vaccines, 78, 107, 110, 114,
      118–19
severity, 1–2, 4, 6, 10–11, 14, 18–21,
   29, 31–2, 35–46, 49–50, 54–6, 58,
   62–3, 71–4, 78–9, 81–3, 85–6,
   89–90, 92–101, 107, 113–14, 120,

severity – *continued*
122, 124, 127, 135, 140, 143, 145–50, 153, 159, 173, 178, 191–2, 194, 197–8, 200, 206–8, 214
smallpox, 107–9, 175
social construction, 3, 144, 149, 209
social distancing, 108, 125
societal norms and values, 33–4
sociology of risk, 32–5
sovereignty, 176, 181, 191
Spanish influenza (1918/1919), 2, 17–21, 58–9, 144–6, 149, 175
stakeholders, 1, 111, 151, 155, 168, 180, 182, 184–5, 189, 195, 204, 207
state actors, 176, 180, 187, 191, 194
the state, role of, 176, 190–1
state sovereignty, 181, 191
structuring of public health, 172–4, 176, 179, 181, 186
summary of pandemic alert phases, 210–13
surveillance, 1, 47–8, 61, 88, 90, 123, 130–2, 137–9, 181, 210, 213
swine flu, *see* H1N1

technology, 19, 116, 119
timeline of events/summary, 210–13
trade, 25, 173, 178
transparency, 136, 151–2, 155, 164–7, 169–70, 172–3, 188, 202, 211
travel restrictions, 125–7, 129
treatment, 43, 123, 146, 199
trust, risk and, 148–52

uncertainty
 evolving threat and, 51–7, 129–31
 future, 51–7, 83–92
 and need for information, 45–51
 and pandemic phases, 74–83
 post-pandemic declaration and, 83–92
 risk and, 31–63, 75, 87, 116, 143, 146–8, 214

scientific, 4–5, 11, 31–63, 80, 103, 106, 122–3, 133, 144, 150, 160, 204, 206, 209, 214
statistical, 57–63, 206
utility and efficacy
 of antivirals, 102, 123–5
 of vaccines, 102–3, 106–9, 131, 160–1

vaccines/vaccination, 102–32
 access and availability, 106, 111–14, 157, 193, 199–200
 adverse reactions, 115, 118, 120–3, 134, 149, 161–3
 cancellation of non-delivered, 158
 carcinogenic nature of, 162–3
 criticism of, 134, 139, 156–63
 degree of purity, 121
 development, 106
 manufacturing industry, 78, 102, 109–15; capacity, 109–10, 114–15; cost/payment, 113, 159–60; and distribution, 2, 106, 110–11, 118; influence/role, 111–15, 135, 142, 156–63, 167–9; interrelations with multiple actors, 111–15, 166–7; novel methods of production, 162–3; as a profit-making enterprise, 113–14, 156; regulatory measures, 118; timing of production, 110–11; vaccine access/distribution, 2, 106, 110–14, 118
 mass, 5, 103, 106–9, 115, 117, 119–20, 122, 131, 139, 159, 175, 183, 199, 207
 post-pandemic period, 88
 profit-making interventions, 102, 113, 156
 quality of, 117, 121, 160
 safety, 102, 107–8, 115–19, 131; clinical trials, 116–17, 163; market testing, 118; regulatory measures, 116–18
 sales of, 158
 stockpiling, 113, 199

utility and efficacy, 35, 102–3,
    106–9, 131, 160–1
WHO's historical successes, 103,
    106–9, 131
  see also management of H1N1
vertical disease campaigns, 108–9, 177
vested interest, 157–8
vigilance, 1, 17–18, 56, 86, 88, 90
  see also monitoring and surveillance

virulence, 19–20, 99
vulnerability, 3, 9, 14, 29, 31, 63–4,
    74, 99, 171, 173, 178, 187, 192,
    197–8, 200, 204

'working together' against the virus,
    187–8

zoonosis, 22–7

GPSR Compliance

The European Union's (EU) General Product Safety Regulation (GPSR) is a set of rules that requires consumer products to be safe and our obligations to ensure this.

If you have any concerns about our products, you can contact us on

ProductSafety@springernature.com

In case Publisher is established outside the EU, the EU authorized representative is:

Springer Nature Customer Service Center GmbH
Europaplatz 3
69115 Heidelberg, Germany

www.ingramcontent.com/pod-product-compliance
Lightning Source LLC
Chambersburg PA
CBHW071615100426
42873CB00004B/48